Excited State
Lifetime Measurements

Excited State Lifetime Measurements

J. N. DEMAS

Department of Chemistry
University of Virginia
Charlottesville, Virginia

1983

ACADEMIC PRESS

A Subsidiary of Harcourt Brace Jovanovich, Publishers

New York   London
Paris   San Diego   San Francisco   São Paulo   Sydney   Tokyo   Toronto

ACADEMIC PRESS, INC.
111 Fifth Avenue, New York, New York 10003

United Kingdom Edition published by
ACADEMIC PRESS, INC. (LONDON) LTD.
24/28 Oval Road, London NW1   7DX

Library of Congress Cataloging in Publication Data

Demas, J. N.
    Excited state lifetime measurements

    Based on a series of lectures presented at the
University of New Mexico, spring, 1967.
    Bibliography: p.
    Includes index.
    1. Excited state chemistry. I. Title. [DNLM:
1. Biophysics. QT 34 D372e]
QD461.5.D45   1982   541.2'8   82-16253
ISBN 0-12-208920-0

PRINTED IN THE UNITED STATES OF AMERICA

83 84 85 86   9 8 7 6 5 4 3 2 1

To David and Stacy

Contents

Chapter 3   Simple Systems

Chapter 4   More Complex Systems

Chapter 5   Least Squares Data Reduction

Chapter 6   Convolution Integrals

Chapter 7   Real Detection Systems (and Does It Matter?)

Chapter 8   Deconvolution Methods

Chapter 9   Experimental Methods

Chapter 10   Special Error Sources

Chapter 11   Testing and Evaluation of Methods and Instruments

Appendix A   Solution of Generalized Response [Eq. (4–2)]

Appendix B   Solution of the Phase Shift Formula [Eq. (4-12)]

Appendix C   Solution of Coupled Equilibria [Eq. (4-25)]

Preface

Excited state lifetime measurements are regularly used to make spectroscopic state assignments, to obtain rate constants for slow and diffusion-limited bimolecular solution processes, to evaluate solid state energy-transfer parameters, to calculate laser thresholds for luminescent substances, and to analyze chemical and biological processes. Lifetime measurements also provide static and dynamic conformational information about small molecules and macromolecules in solutions, micelles, membranes, and monolayers. The results of lifetime measurements are reported in journals of biology, physics, chemistry, spectroscopy, and analytical methods, demonstrating their wide acceptance as tools of the physical and biological sciences.

Unfortunately, although there is a wealth of information applicable to lifetime measurements, it is scattered and poorly cross-referenced. Indeed, many of the necessary mathematical formulations are to be found under such seemingly unrelated fields as chemical kinetics, radiochemistry, computer science, and electrical circuit theory.

This book, which unifies and clarifies much of this information, began as a series of lectures presented in the spring of 1967 to the research group of Professor G. A. Crosby at the University of New Mexico. Afterward, the material having been used and revised for the introduction of new undergraduate and graduate students and postdoctorals to the area of excited state lifetime measurements, this book was written.

The significance of excited state lifetimes as well as the methods of measuring them and the data treatment are described in this volume. It is intended for analytical chemists, photochemists, photobiologists, spectroscopists, and physicists. It is especially suitable for the novice, but the more knowledgeable reader is likely to find much new material also. Only elementary calculus and differential equations are required; however, the reader is carried through sophisticated state-of-the-art data-reduction techniques. Real systems are described early to permit the novice to foresee the practical utilization of lifetime measurements. Adequate descriptions of different experimental techniques are given to permit the reader to judge their relative merits. The goal is a self-contained presentation of the experimental evaluation of the parameters of excited state kinetics with satisfying rigor but without undue complexity. Above all, the book is intended to be practical and complete enough to be used by new students for independent study.

Chapter 1 is a brief elementary discussion of the applications of excited state lifetime measurements. Although nonrigorous, it illustrates a variety of practical and fundamental applications of lifetime measurements. Chapter 2 discusses a number of experimental methods for measuring excited state lifetimes. This aids the uninitiated in visualizing the experimental measurements undertaken and their relationship to the subsequent mathematical developments. Chapter 3 describes "simple systems," focusing principally on first-order or pseudo-first-order processes, and discusses a variety of interesting chemical systems that give rise to these processes.

Chapter 4 treats a series of appreciably more complex systems, such as serial decay kinetics, the steady state approximation, resonance energy transfer, and excited state acid–base and excimer reactions. The convolution problem, where the decay times of the system are comparable to the duration of the excitation pulse, is first introduced here. Chapter 5 is a mathematical digression on the method of least squares fitting and its uses and misuses.

Chapter 6 introduces the impulse-response concept and the powerful convolution integral and its applications. Frequently, the convolution approach replaces pages of laborious, sometimes obscure, kinetic analysis with a clear solution of only a few lines. In Chapter 7 the real detector systems are analyzed using the convolution integral. Chapter 8 describes deconvolution methods. Chapter 9 provides more details of instrumentation. This chapter is not exhaustive but does expand on Chapter 2 and includes some state-of-the-art methods as well as descriptions of some frequently overlooked design criteria in more conventional systems. Chapter 10 describes insidious and special error sources. Finally, Chapter

11 describes approaches, especially digital simulation, to the evaluation of new methodologies. Indeed, most of the examples in this book were derived by digital simulation.

Some long or nasty mathematical derivations are relegated to the appendixes. This prevents their disrupting the text, permits verbal description of the phenomena, and avoids boring the more knowledgeable reader. Several useful computer programs for data reduction and simulations are included. The language selected was microcomputer BASIC. This choice reflects the accelerating trend toward dedicated laboratory mini- and microcomputers and their incorporation in commercial instruments. It is also consistent with the reduced role of the dying dinosaur—the big central mainframe computer.

At various points in this book we discuss specific systems and technology. In such a rapidly changing area, some material will be quickly dated but will remain a useful reference point for comparison with recent developments. Although more modern instrumentation will give improved performance, older but much less costly instrumentation will continue in wide use. Further, many of the concepts will remain intact, and the approaches used to solve past problems should be quite instructive.

I am indebted to my wife who typed and edited her way through this manuscript and to my children, David and Stacy, for the work they did on this project. I am very grateful to the reviewers and to Ben Hauenstein for many helpful suggestions, and I also thank Scott Buell for his help. A special thanks to my research group who suffered patiently through all of those canceled group meetings.

CHAPTER

1

Applications

A. INTRODUCTION

The measurement of excited state lifetimes has become a pervasive and invaluable tool in the realms of experimental and theoretical spectroscopy, solution kinetics, solid state physics, energy transfer, laser and solar energy technology, analytical chemistry, and biology. Lifetimes are regularly used to make state assignments, to obtain quenching constants for slow and diffusion-limited bimolecular solution processes, and to evaluate solid state energy transfer parameters. They also find application in evaluating laser dyes, developing analytical methods, and determining static and dynamic conformational information about macromolecules, micelles, membranes, and monolayers. The results of lifetime measurements are reported in journals of biology, physics, chemistry, spectroscopy, and analysis. In short, luminescence lifetime measurements have become an indispensable tool in the physical and biological sciences. As an example of this importance, a two-week NATO Advanced Study Institute on time-resolved fluorescence spectroscopy in biochemistry and biology was held in Saint Andrews, Scotland, in 1980. These proceedings (Cundall and Dale, 1983) make excellent reading for anyone interested in the role of lifetime measurements.

There is a great wealth of information in the literature that is applicable to lifetime measurements. Unfortunately, this information is widely

1

scattered and frequently poorly cross-referenced. In addition, many of the mathematical formulations necessary for lifetime work are to be found under such seemingly unrelated subjects as chemical kinetics, radiochemistry, and electrical circuit theory. In an attempt to help clarify these problems, this book presents a broad range of topics that will be of interest to workers in a variety of fields and to newcomers trying to learn these concepts from the scientific literature. An attempt has been made to lay the necessary mathematical groundwork and to provide a range of applications that will whet the appetite of the reader for a working knowledge of the mathematical and experimental techniques of lifetime measurements. The goal has been to provide reasonable rigor without overwhelming mathematical complexity; a familiarity with elementary calculus and a slight knowledge of differential equations and some kinetics are assumed. In an attempt to make the underlying principles clearer, more tedious mathematical developments have been relegated to the appendixes.

The remainder of this chapter is devoted to a brief discussion of the applications of excited state lifetime measurements and a summary of the abbreviations used throughout the book. No attempt is made to discuss basic spectroscopy. Readers unfamiliar with such concepts as absorption, emission, and singlet and triplet states should consult introductory physical chemistry texts as well as the books of Turro (1978) and Birks (1970). While the treatment here is terse and informal, it provides the reader with a view of the pervasiveness of these measurements and the variety of practical and fundamental problems to which they can be applied.

B. ANALYTICAL CHEMISTRY

Very selective and sensitive methods of microanalysis use steady state fluorescence or phosphorescence (Parker, 1968). Unlike traditional absorption measurements, the sample optical densities can be exceedingly low. Further, the excitation and emission wavelengths can be varied to optimize sensitivity and to discriminate against other luminescent components in a mixture. The combined luminescence and excitation spectra provide a valuable identification tool.

Many times, however, luminescence spectra are not adequate for unambiguous identification. Fluorescence and phosphorescence lifetimes provide another diagnostic tool since, frequently, materials having very similar spectra can have greatly different lifetimes. In the analysis of mixtures, differences in lifetimes can be exploited to separate and quantitate different species. For example, if the components of a mixture exhibit different lifetimes, when the sample is excited with a short excitation flash, the observed decay usually will be a sum of the exponential decays contributed by all the components. By mathematically analyzing the composite decay curve, the individual lifetimes and the contributions of each component to the curve can be determined. This may permit simultaneous analysis of two or more component mixtures having virtually identical absorption and emission spectra; they need only have different lifetimes. Thus, time resolution adds an additional degree of freedom to normal luminescence analysis. Further, by analyzing the decay curves as a function of excitation and emission wavelength, even more information about the sample components can be obtained (Winefordner, 1973; Johnson *et al.*, 1977; Love and Upton, 1980). There are, of course, numerous conditions that must be satisfied, but in the very dilute samples encountered in analyses these conditions are frequently met.

Traditionally, time-resolved luminescence spectra were limited to low-temperature phosphorescence decays with lifetimes of hundreds of microseconds to seconds. More recently, the room temperature phosphorescence of a variety of organic species on filter paper has been observed and used for quantitative microanalysis (see, e.g., Goeringer and Pardue, 1979). Advances in single-photon counting decay equipment, mathematical deconvolution techniques, and low-cost on-line computers have, however, made time resolution possible in the low-nanosecond region even when the sample lifetimes are comparable to the excitation pulse width. A study of the analytical possibilities and limitations of such deconvolution methods has recently been made by Love and Upton (1980). With an excitation pulse having a 1.3-nsec decaying tail, these authors resolved a sample containing two components with lifetimes of 19.4 and 2.7 nsec as long as the intensity ratio was 1/15 or larger. Using digital simulations they concluded that about an order of magnitude improvement was feasible over their experimental results.

Small and Isenberg (1977, 1983) have shown that up to three exponential decays can be successfully resolved with remarkable precision and accuracy. The lifetimes must be well separated, and the ratio of the intensities cannot be varied too widely.

C. MOLECULAR YARDSTICK

A very bothersome problem of protein and polymer chemistry is that of identifying the distances between specific sites or molecules. X-ray determinations of proteins are powerful but frequently inappropriate for the many proteins that do not crystallize. Further, crystal structure is not necessarily the same as solution structure. A powerful alternative tool uses molecular luminescence. A luminescent probe is attached at one site and a quencher (excited state deactivator) at the other or, alternatively, one attaches a quencher and monitors an intrinsic luminescent component. The quencher is selected so that it deactivates the luminescent species by a long-range resonance (Förster) energy transfer. By measuring the rate of deactivation of the probe, and from spectral knowledge of the probe and deactivator, the distance between the two sites can be estimated (Stryer and Haugland, 1967; Stryer, 1978; Birks, 1970; Bennett, 1964; Zimmerman *et al.*, 1980; Lakowicz, 1983).

This approach has been generalized to when the probe and quencher are attached to specific sites on a molecule (e.g., a polymer) that is not rigid and undergoes appreciable dynamic motion. Under these conditions, not only can the average distance be estimated but the size and distribution of the molecular motions about the equilibrium separation can be calculated (Steinberg and Katchalski, 1968; Steinberg *et al.*, 1983).

D. MOLECULAR DYNAMICS

Again, in protein, membrane, micelle, and polymer chemistry it is important to know how rapidly certain portions of the molecule can rotate. To monitor this motion, fluorescent probes are attached to the portion in question. Then the probe is excited with a short polarized excitation pulse, and the rates of decay of the probe emissions that are polarized parallel and perpendicular to the exciting light are monitored. From the rates of decay of the two polarizations, it is possible to estimate how rapidly the probe rotates in different molecular directions. If the probe is rigidly attached to a portion of the protein, the probe tracks the motion of this portion, which determines the polarized properties of the decay (Spencer and Weber, 1970; Wahl, 1975, 1983; Tao, 1969; Lakowicz, 1983).

The dynamics of interactions of chromophores on synthetic polymers can also be studied by their decay characteristics. Many organic chromophores will not interact with each other in the ground state, but two will form an excited state complex (exciplex) with each other if one is excited. This exciplex has luminescence characteristics that differ from those of the excited single molecule. By using a short pulse excitation and watching the grow-in and decay of the exciplex relative to the monomer, dynamic motions of the polymer chain motion can be estimated (Beaven *et al.*, 1979).

E. MOLECULAR STRUCTURE

In protein chemistry, knowledge of the environment of metal ions at binding sites is important in understanding such features as enzyme mechanisms. Luminescent probes provide useful information in this area. Ions such as Tb^{3+} and Eu^{3+} luminesce, and their lifetimes are very sensitive to their chemical surroundings, especially hydration (Kropp and Windsor, 1963, 1965, 1966). These ions will replace Ca^{2+} and Mg^{2+} in a variety of proteins. Then, by monitoring their lifetimes, the hydration of or other changes in the ion environment can be ascertained as the protein or its environment is modified (Horrocks and Sudnick, 1979).

Hauenstein *et al.* (1983b) have extended the use of the deuterium isotope effect as a probe of the solvent environment around transition metal sensitizers bound to organized media such as micelles. Their approach is likely to prove useful in studies of membranes, monolayers, and chemically modified electrodes.

Lifetimes of rare earths are also useful as probes of the structure of various complexes in solution. Brittain (1979) has combined lifetime and intensity measurements on mixtures of Eu^{3+} and Tb^{3+} with different ligands to show that at high pH values the complexes exist as aggregates of two or more ions.

Bauer *et al.* (1982a,b) have used lifetime measurements to probe the binding and motions of organic molecules bound to silica gel, porous Vycor, alumina, and calcium fluoride. Photophysics and photochemistry of these systems were examined.

F. THERMOMETERS AND MISCELLANEOUS ITEMS

Sholes and Small (1980) have described the use of a ruby chip as a suitable thermometer for in vivo temperature measurements. Such an approach is useful where rf or microwave heating therapy for malignant tumors is used and the thermometer must operate in a hostile electromagnetic environment without any electrical connections. Over the temperature range 35–47°C, the lifetime of the ruby chip decreased from 3.49 msec to 3.38 msec. With an analog sampling procedure, a camera shutter, a silicon photodiode, and a 12-bit analog-to-digital converter (ADC), the system was accurate to ±0.3°C over this range, and the accuracy was limited only by the resolution of the ADC.

Crosby and co-workers (Harrigan and Crosby, 1973) have reported dramatic lifetime changes in ruthenium(II) complexes dispersed in plexiglass at temperatures below 20°K. Such systems show considerable promise as cryogenic thermometers, although they were not used as such.

Gibson and Rest (1980) have used lifetime measurements to determine the refractive indexes of low-temperature rare gas matrices. These measurements would be exceptionally difficult with any other approach.

Cramer and Spears (1978) have used the fluorescence lifetimes of rose bengal (77–2570 psec) in a variety of different solvents to assess the extent of hydrogen bonding in the dye. They interpreted their results on the basis of increased singlet–triplet splitting with decreased hydrogen bonding, which reduced the rate of intersystem crossing from the emitting singlet to the triplet state.

G. ASSIGNMENT OF EXCITED STATE TYPES

In many cases the nature of the emitting states cannot be determined just from the spectral distribution of the emission. But the lifetime can be used to differentiate between spin-allowed fluorescences, with lifetimes typically less than 100 nsec, and spin-forbidden phosphorescences, with lifetimes greater than 10 μsec (under conditions where the luminescence yield is reasonably high). In more difficult cases the charge transfer (CT) and ligand-localized luminescences of Ru(II) and Ir(III) complexes have been assigned as phosphorescences on the basis of luminescence yields and lifetimes. See, for example, the reviews by DeArmond (1974) and by Crosby (1975). Recently, however, a more detailed theoretical, experimental analysis indicates that the CT emissions arise from spin orbit states rather than from triplets or singlets (Hipps and Crosby, 1975; Elfring and Crosby, 1981).

As another example, Carter and Gillispie (1980) have used lifetime arguments to assign the emission of dichromate ion pairs to a spin-forbidden phosphorescence. In the analytical area, the very long lifetimes (multiseconds) of emissions of aromatic molecules dispersed on filter paper at room temperature have been used to assign them unambiguously to room temperature phosphorescences (Goeringer and Pardue, 1979).

Peterson and Demas (1979) have used lifetime measurements to demonstrate excited state inversion on protonation of a metal complex. They showed that the lowest excited state of Ru(phen)$_2$(CN)$_2$ (phen = 1,10-phenanthroline) changes from a CT state to a $^3(\pi-\pi^*)$ ligand-localized state on protonation of the cyanides. This assignment was possible because of the characteristic lifetime of CT emissions and of ligand-localized phosphorescences.

Krug and Demas (1979) have used lifetime measurements to assign the emission [Ru(bt)$_3$]$^{2+}$ (bt = 2,2'-bi-2-thiazoline) to a CT emission. This complex is unique, since bt contains only an α-diimine moiety rather than the full ring aromatized system characteristic of other Ru(II) CT emitters.

H. MICELLE COMPOSITION

Tris(2,2'-bipyridine)ruthenium(II), [Ru(bpy)]$^{2+}$, has been widely used as a photosensitizer in both homogeneous and heterogeneous systems (Balzani et al., 1975). In micelles, abnormal excited state decay kinetics have been used to show that [Ru(bpy)$_3$]$^{2+}$ forms aggregates with sodium dodecyl sulfate at concentrations below the critical micelle concentration and, further, that these aggregates contain two or more [Ru(bpy)$_3$]$^{2+}$ molecules per aggregate (Baxendale and Rodgers, 1980).

Rodgers (1981) used picosecond luminescence measurements to study the interactions of rose bengal with micelles. At high concentrations of dye, multiple occupancy of the micelle by rose bengal appeared to lead to singlet–singlet annihilation.

I. EXCITED STATE ACID–BASE REACTIONS AND TAUTOMERIZATION

In the excited state, many molecules can greatly change their acid–base properties. For example, 2-naphthol is about six orders of magnitude more acidic in the excited state than in the ground state (Parker, 1968). Detailed analyses of luminescence spectra or, better yet, time-resolved emission spectra permit measurement of the excited state pK values as well as the rates of the proton transfer reactions. See, for example, the detailed analysis of the naphthol system by Laws and Brand (1979) and by Harris and Selinger (1980). Ultrafast proton transfer reactions have also been observed in reverse micelles (Escabi-Perez and Fendler, 1978).

Such excited state acid–base properties are not limited to organic molecules. Peterson and Demas (1976, 1979) have demonstrated that the protonated forms of [Ru(L)$_2$(CN)$_2$] (L = 2,2'-bipyridine or 1,10-phenanthroline) are about five orders of magnitude more acidic in the excited state than in the ground state.

In molecules containing more than one acid–base function a protonated and an unprotonated site may exist. On excitation the unprotonated site may become more acidic than the protonated one. Under these conditions an intramolecular proton transfer can occur in the excited state to yield a tautomer. By means of time-resolved spectroscopy, the rates and mechanisms of this transfer can be studied. For example, Choi *et al.* (1980) studied the phototautomerization of lumichrome (7,8-dimethylalloxazine). They found that a proton transfer catalyst greatly accelerated the transfer and that efficient transfer occurred at temperatures higher than 100°K. Watts and Bergeron (1978) have reported evidence for phototautomerization in a novel Ir(III) complex.

J. SOLVENT RELAXATION

On excitation, a molecule may no longer be in its most stable configuration. Because of electronic rearrangement, and especially because of changes in dipole moments, the organization of solvent molecules around the ground state molecule is not the most stable one for the excited state. If the excited state lifetime is long enough, the solvent and the excited molecules can rearrange and acquire a more stable configuration for the excited state. DeToma (1983) and Lakowicz (1983) have reviewed many of the theories in this area.

If the solvent reorganization time is exceedingly short, virtually all emission occurs from the thermally equilibrated state. If the solvent relaxation time is long compared to the excited state lifetime, then all the emission will arise from the initially excited conformation which has the ground state solvent configuration. If, however, the two relaxation times are comparable, the emitting species will change with time following a short excitation pulse. If the emission spectra of the different forms vary significantly, then the emission spectrum of the sample will vary with the delay following the pulsed excitation. At short delay times the spectrum will match the initially excited form, while at long times the emission spectrum will approach that of the thermally equilibrated molecule. Emission spectra taken at different delay times are called time-resolved emission spectra.

Solvent relaxation is, in fact, a complex function of solvent viscosity, polarity, polarization, and specific excited state interactions. It now appears certain that the transition from the initial to the final conformation occurs in at least several small steps, and that the observed emission cannot be represented as a sum of two emissions but must, in fact, be treated as a series or even a continuum of emission spectra that vary with time (DeToma, 1983; Lakowicz, 1983).

Since relaxation times are dependent on the solvent viscosity, it is possible to change the emission properties of a system merely by varying the temperature, which changes the solvent viscosity. Thus, at low temperatures the solvent relaxation is slowed and the emission arises exclusively from the initially excited form. At high temperatures the emission is exclusively from the equilibrated form. At intermediate temperatures only partial equilibrium can occur, and strongly time-dependent time-resolved emission spectra result. Such data can yield significant information about solvent–solute interactions and excited state molecular dynamics.

The existence of unappreciated solvent relaxation effects can have a profound and deleterious effect on data interpretation. Low-temperature luminescence spectra are used for estimating excited state energies of photosensitizers, since the emissions are generally much sharper and brighter than under the fluid room temperature conditions where they are used as photosensitizers. Indeed, many triplet photosensitizers do not emit at room temperature. Thus, if there are large solvent reorganization energies, the equilibrated state responsible for room temperature sensitizations will have a much lower energy than the low-temperature unequilibrated one.

Similarly, the use of fluorescent probes to measure the environmental polarity of binding sites on proteins can likewise encounter difficulty. Most polarity probes exhibit large solvent relaxation energies. If the solvent relaxation time of the environment at the binding site is much slower than that for the solvents used to calibrate the probe, the emission will largely arise from the unequilibrated form and large errors in the polarity of the environment will result. Estimates of rotational times by lifetime measurements or even simple decay time measurements are modulated by these parasitic decays, which enormously increases the complexity of interpretation (Lakowicz, 1983).

K. SOLAR ENERGY CONVERSION

Studies on the quenching constants of luminescent Ru(II) complexes are an integral part of studying the bimolecular excited state electron transfer reactions of these systems for use in photogalvanic cells. For example, Lin and Sutin (1976) have studied quenching by Fe^{3+}, and De-Graff and Demas (1980) have studied quenching by Hg^{2+} in the evaluation of photogalvanic properties.

Picosecond spectroscopy has been used to study the excited state properties of Ru(II) and Fe(II) α-diimine complexes and to assess their excited state electron transfer properties and thus potential utilization as photon-harvesting antenna in conversion schemes (Creutz et al., 1980). The results demonstrated that the excited state reducing power of the Fe(II) complexes was too low and the lifetimes too short for them to be very promising photosensitizers.

Fundamental studies on the photosynthetic process routinely use conventional and picosecond luminescence and flash photolysis for monitoring energy migration and electron transfer events in chloroplasts (Moya, 1983). The numerous studies on photodecomposition of water also de-

TABLE 1-1

Symbols Used in the Text

Symbol	Definition
a	Least squares fitting parameters
$d(t)$	True decay curve
$D(t)$	Observed decay curve
$e(t)$	True excitation profile
$E(t)$	Observed excitation profile
ε	Molar extinction coefficient
f	Frequency
$i(t)$	Sample impulse response
k	Rate constant
K	Preexponential factor
m_k	kth moment of $E(t)$
μ_k	kth moment of $D(t)$
ω	Angular frequency $= 2\pi f$
ϕ	Luminescence intensity
Q	Luminescence quantum yield
$R(t)$	Observed decay of reference molecule
R_i, R_{wi}	Residual and weighted residuals
σ	Standard deviation or radius of gyration
t	Time
$T_{1/2}$	Half-life
τ	Mean lifetime
τ_r	Radiative lifetime
τ_0	Unquenched lifetime
v	Number of degrees of freedom
χ^2, χ_r^2	Statistical chi-square and reduced chi-square

pend on knowledge of excited state lifetimes and interactions with other molecules (Kalyanasundaram, 1982).

Kelder and Rabani (1981) have attempted energy pooling of multiple photons by concentrating absorbing acceptors on polyelectrolytes. Using excited state decays, they demonstrated interactions of two or more excited states, but pooling attempts failed.

L. CONVENTIONS

An attempt has been made to utilize a consistent set of symbols throughout this book. Where possible, the standard definitions used in the literature have been employed. Table 1-1 summarizes these symbols.

One major deviation in symbols, however, is for excitation and decay curves. In the literature $F(t)$ is widely used to stand for the fluorescence decay curve. Since the analyses presented here apply to phosphorescence and to the spin orbit emissions the author works with, $D(t)$ is used to represent the decay curve. As a mnemonic aid $E(t)$ is used to stand for the flash excitation curve.

Another convention involves the time dependence of excitation and decay curves. Unless explicitly stated otherwise, $E(t)$ and $D(t)$ are assumed to be zero for all times before $t = 0$.

2

Methods of Measuring Lifetimes

A. INTRODUCTION

We now turn to methods of measuring excited state lifetimes. The purpose is to provide the unfamiliar reader with the necessary background to appreciate how the excited state concentrations referred to in the next chapters can actually be followed. An incredible number of different approaches have been developed, and our discussion is thus limited. More experimental information is supplied in Chapter 9.

We begin with luminescence approaches where the excited state concentrations are followed by their luminescence; these include flash and phase shift measurements. We then cover briefly flash methods employing absorption monitoring of excited state concentrations. A brief discussion of the very expensive and difficult, but extremely important, picosecond methods follows. Finally, a number of miscellaneous methods are described.

B. LUMINESCENCE METHODS

The most common methods of measuring excited state lifetimes involve monitoring the time dependence of the sample emission while exciting it with pulsed or modulated excitation. From the sample response

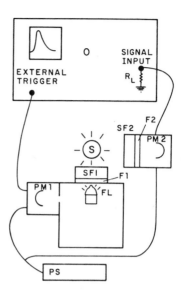

Fig. 2-1. Typical flash-excited decay time instrument. FL, Flash lamp and control electronics; PM1, trigger phototube; PM2, signal phototube; PS, high-voltage power supply; F1, SF1, excitation filters; S, sample; SF2, F2, emission filters; O, oscilloscope or transient recorder; R_L, signal load resistor. [Reprinted with permission from Demas (1976). Copyright 1976 by the Division of Chemical Education, American Chemical Society.]

relative to the excitation pulse, information about the decay kinetics can be obtained. Figure 2-1 shows a typical flash apparatus. With variations in the excitation source, detection device, and display, it is representative of most luminescence decay time instruments. Generally, photomultipliers (PMTs) are used to measure luminescence, but if exceptionally intense laser sources are used, photodiodes may be adequate. We describe here a variety of recording techniques that are used and some of the advantages and limitations of each. Detailed data treatment, theoretical developments, and more specialized instrument descriptions will be reserved for later chapters.

1. Strip Chart Recorders

For long lifetimes (>1 sec) a normal mechanical strip chart recorder is frequently adequate. Typically, recorders have response times of ~0.5 sec for a full scale step, and some respond in 0.1 sec. Thus they are adequate for the slow phosphorescence of many organic molecules. Oscillographic recorders with photosensitive papers extend this range by perhaps an order of magnitude, but their high cost generally makes them unattractive.

Logarithmic amplifiers can greatly improve the range and effective response time of a recorder. Consider a 10-V peak signal with an exponential lifetime of 1 sec that gives a decay of 10 V $\exp(-t)$, where t is in seconds. The maximum rate of change in this signal, dV/dt, at $t = 0$ is -10

V/sec. If, however, the logarithm is taken with an amplifier whose output is 1 V/decade of signal change, then the maximum rate of change is −0.4 V/sec. Thus, not only is the maximum rate of change reduced, but the dynamic range and readability of the waveforms are greatly enhanced.

2. Oscilloscopes

Until recently, the standard data collection method employed an oscilloscope coupled with a camera. Although now replaced in many laboratories by single-photon counting (SPC) instruments and transient recorders, oscilloscopes still represent the workhorse device for many researchers.

The oscilloscope has many merits. It is easy to set up, and for phenomena that are not too short-lived the image is easily viewed, thus facilitating sample alignment, phototube voltage, etc. Indeed, even with other types of data acquisition, a conventional oscilloscope is invaluable for system setup and debugging. Oscilloscopes are, however, generally tedious to use. Accuracy is modest, which limits the accuracy of data processing performed on digitized data.

Oscilloscopes can be used for data acquisition up to the 5–10-sec range, although a strip chart recorder is more appropriate. The high-speed limit depends on the oscilloscope. Oscilloscopes are rated by their high-frequency cutoff f_c, typically in megahertz. The rise times of the oscilloscope can roughly be related to f_c by

$$t_r \, (\mu \, \sec) = \frac{1}{3f_c} \, (\text{MHz}) \tag{2-1}$$

Thus, a 15-MHz scope has a 22-nsec rise time, a 50-MHz scope a 7-nsec rise time, and a 250-MHz scope a 1.3-nsec rise time. A 1-GHz bandwidth sampling oscilloscope (vide infra) has an 0.3-nsec rise time.

Rise time is not the whole story. The intensity of the image produced per pass of the electron beam across the screen depends on the phosphor and the accelerating potential used on the cathode ray tube (CRT). Many fast oscilloscopes are unsuitable for viewing low-repetition-rate signals because the acclerating potential is low. For high-repetition-rate signals there is no problem, because the beam passes over the screen many times during the integrating period of the eye. With older oscilloscopes, it was extremely difficult to view low-repetition-rate signals even in the 20 nsec/cm range. Many new very high-speed oscilloscopes can produce viewable single-shot transients even in the 1–5 nsec/cm range.

A variety of memory oscilloscopes are also available. These scopes produce a stored semipermanent image on the screen even with a single shot. In the past most of these memory scopes were relatively slow. For a single transient, the sweep typically had to be 0.2–10 μsec/cm or slower to produce a usable trace. Here again technology has advanced to single-shot writing speeds of 2000–2500 cm/μsec (Tektronix 7834 and Hewlett-Packard 1727A), which permits the recording of 8-cm peak-to-peak 100-MHz sine waves single shot. For a review of storage oscilloscope technology and availability see Nelson (1981).

It is not always appreciated, but with repetitive signals a memory scope can usually be used at speeds 10–100 times the single-shot writing speed. The memory screen responds to an accumulated electron dose. Thus, by writing 10 pulses on the screen an image comparable to a single shot at $\frac{1}{10}$ the speed can be obtained. In practice, the situation is not quite this good. The memory system is not completely linear, and more pulses are usually required. Also, the steady leakage of electrons onto the screen slowly builds up a fog if it is left in the write mode too long. Further, if the signal is noisy, the screen will not be written on at exactly the same place each time, and a much longer time will be required to build up the image. This last point can be used quite advantageously as a poor man's signal averager. On the average the signal will occur more often at the mean than at the extremes. Thus, the center of the signal will be written on more often than the noisy extremes. For a relatively high rate signal, a high degree of signal averaging in the data acquisition process is possible.

Generally, most oscillographic data are photographed. Polaroid film is almost indispensable for this, so that the operator can decide immediately whether a valid recording has been obtained. Such film comes in a variety of types, the standard being the ASA 3000 speed which is adequate for all but the highest speeds. An ASA 1000 speed is also available.

A memory oscilloscope is invaluable in keeping film costs down. One creates what is visually an acceptable viewing image and then photographs it. Once the necessary conditions for photographing such an image are established, they are independent of whether the generated display arises from a single-shot slow waveform or many averages of a fast one.

The author has found the slow Polaroid type-46L transparency slide film to be exceptionally useful for work with a memory oscilloscope. After development and drying (it does not have to be fixed unless archival permanence is required), it can be put between two glass plates, projected with a photographic enlarger onto graph paper, and traced. Except for scaling, the data can then be read directly off of the graph paper. Figure 2-2 is a representative photograph taken using a laser excitation system. The human eye, incidentally, is an excellent signal averager for data of the form shown in Fig. 2-2. It is easy to draw a line very accurately down the

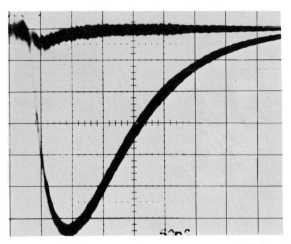

Fig. 2-2. Typical oscilloscope photograph. In this case each division is 50 nsec. The decay is a laser-excited luminescence of $Os(\phi_2phen)_3^{2+}$ (ϕ_2phen = 4,7-diphenyl-1,10-phenanthroline). Increasing intensity is downward, and a sample blank is shown for comparison.

center of the trace; the reproducibility of lifetimes measured with such photographs is typically 1–3%.

Another trick for digitizing that the author's group has used employs a plotter such as a Helwett-Packard 9872A. This plotter has a digitizing sight which, when controlled by a calculator, can function as a digitizer. The data are read directly into, and can be processed by, the calculator. As an added advantage, the plotter function is not usurped by the digitization. It is thus possible to move the digitization sight automatically in a specified number of discrete steps across the waveform. With only slightly increased complexity, a simple predictor can be added to the program so that the sight moves to the expected point. With most relatively smooth curves the predicting is so good that only minor trimming, or none at all, is required for the final positioning. With this system it is possible to digitize 100 points from a photograph at rates in excess of 1/ sec. Also, commercially available digitizers with a processing computer can now be purchased for less than $3000.

Oscilloscopic data recording has severe limitations. The most significant ones are the limited accuracy (2–5% full scale) and the limited dynamic range (roughly a factor of 5–10). Some workers have tried to circumvent the latter problem by taking more than one photograph with the preamplifier set on progressively higher gains; the early portion of the waveform goes off scale, and the later portion is expanded. Overlapping the sections can be tedious and complex, especially if pulse-to-pulse intensity is not good. Frequently, however, unimproved or worse results are obtained, since oscilloscopes are not designed for this use and the preamplifier can overload on the off-scale portions of the transient. The recovery time can be so long that the transient is complete before the oscilloscope recovers. Worse, however, is that the overload recovery process is sometimes nearly exponential. Thus, the expanded tail of the waveform may just be the overload recovery of the oscilloscope rather than a fundamental process. Those who attempt such expansion techniques should verify that their oscilloscope is indeed recovering in the time available and at the preamplifier settings used for data acquisition.

3. Sampling Methods

Particularly in the past, when ultrahigh-speed oscilloscopes were not available, sampling techniques were used for viewing repetitive fast transients. Instead of trying to view the entire transient on each occurrence, one acquires and views only a single point from the transient each time it occurs. By selecting different points from the transient, a complete representation of the transient can be gradually generated.

The standard sampling device for many years has been the sampling oscilloscope. Figure 2-3 shows a schematic representation. The critical elements are a triggered fast linear time base similar to that of a conventional oscilloscope, a slow-sweep generator, an ultrafast analog comparator, and a sample and hold (S/H). Initially the slow-sweep generator is set at its lowest value. On triggering, the fast-sweep generator starts a linear rise. At the instant this fast signal crosses the value of the slow one, the comparator senses this and strobes the S/H to latch onto a data point. Until a new triggering, the output of the S/H equals the amplitude of the input signal at the instant of crossing. The sampled point is then plotted on the CRT which functions as an $X-Y$ plotter. The amplitude of the slow sweep is used as the X input. Since the delay between triggering and the crossing of the two sweeps is directly proportional to the amplitude of the slow sweep, the X coordinate of the display is directly related to the time of the sampled point on the waveform. After the acquisition of one point, the amplitude of the slow sweep is incremented, which causes a longer delay between triggering and sampling. Thus, the transient is sequentially built up pointwise by sampling at progressively longer times. In a sampling oscilloscope, the density of points (i.e., the interval between samples) can be set, and after the acquisition of one transient the entire process is started from the beginning.

The rise times for slow-sampling oscilloscopes are 0.3–1.0 nsec, and fast ones have 10–100 psec rise times. They are, thus, the devices of choice for repetitive subnanosecond signals. Further, since each point (X,

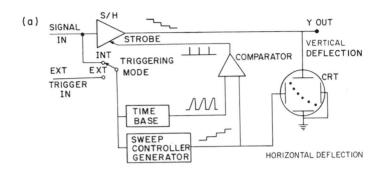

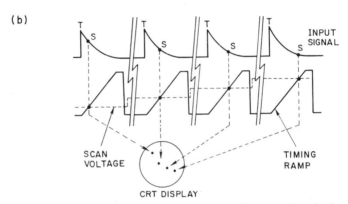

Fig. 2-3. Schematic representation of a sampling oscilloscope (a) and relevant waveforms (b). (a) The time base is triggered on each occurrence of a waveform. The sweep controller output is increased by a fixed amount after each sweep. A CRT display is used as an X–Y point plotter. S/H is triggered once per waveform. (b) T indicates the trigger point at which the linear time base sweep is actuated. The S/H is strobed at the sampling points that occur when the scan voltage (– – –) and linear timing ramp (——) are equal.

Y) is held on the screen for an extended period, there is never the problem with viewing intensity for low-repetition-rate signals that there is with conventional oscilloscopes. Finally, the sampling process lends itself to a very powerful signal-averaging technique called boxcar integration.

In a boxcar integrator, instead of rapidly sweeping across the entire waveform, the slow sweep is increased much more slowly. In this way many points are effectively collected at each time delay. Then, merely by putting a smoothing filter on the output of the S/H, a signal-averaged representation of the waveform can be obtained. Generally, the sweep is done slowly enough so that the output can be displayed on a mechanical recorder. The sampling oscilloscope has formed the basis of many ultra-high-speed boxcar integrators, since all that is necessary is to disconnect the internal sweep generator and replace it with an external one, as well as connect the S/H output signal to a filter.

There are also commercial boxcar integrators that do not use sampling oscilloscope circuitry but are optimized for slower phenomena. Others such as the PAR 162 series can use either oscilloscope sampling circuitry for very fast phenomena or conventional slow boxcar circuitry for slower phenomena.

There are several disadvantages of sampling methods. The linearity and dynamic range are limited. Some workers obtain two or three decades of linearity, but this is critically dependent on the sampling circuitry. Certainly, however, excellent linearity over one or two decades is normal. Also, the signal must be stable and repetitive, which rules out photochemically unstable systems where only a few pulses can produce a significant change in sample composition. Additionally, if the excitation intensity varies over the course of the scan, this source variation translates directly into signal distortions, although a double-beam ratio correction scheme can solve this. Pearson *et al.* (1982) have described a way to circumvent this problem.

In spite of the competition from single-photon counting instruments, the lower cost and more general nature of sampling-based systems will make them viable alternatives in the foreseeable future. Low-cost microcomputerized systems will further enhance their attractiveness.

4. Single-Photon Counting Instruments

One of the most popular approaches for measuring short luminescence lifetimes is the single-photon counting time-correlated method. Unlike virtually all other techniques that depend on there being relatively high emission intensities, SPC depends on the intensity being low. Figure 2-4 shows a simple SPC instrument. Its key parts consist of a flash lamp, a start PMT, a stop PMT, a time-to-amplitude converter (TAC), and a multichannel analyzer (MCA) used in the pulse height analysis (PHA) mode. The TAC generates an output pulse whose amplitude is directly proportional to the time between the start and the stop pulses. In the PHA mode the MCA digitizes the height of the input pulse and increases by one the contents of the memory channel corresponding to this digital value.

In operation, the firing of the flash lamp generates a start pulse to the TAC. The detection of a stop pulse by the sample PMT stops the TAC. The TAC output pulse is presented to the MCA which determines the memory channel corresponding to the delay time between the start and

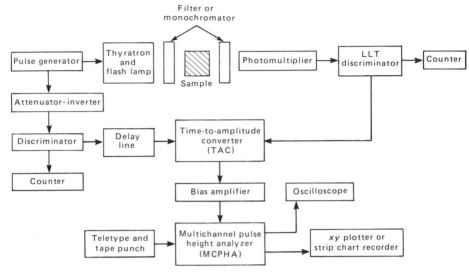

Fig. 2-4. Schematic diagram of a single-photon counting nanosecond fluorimeter. In a modern instrument jitter is reduced by replacing the low-level timing (LLT) discriminator with a constant-fraction discriminator (CFD). The phototube or generator pulses are converted into TAC-compatible signals by the discriminators. [Reprinted from Ware (1971) by courtesy of Marcel Dekker, Inc.]

stop pulses. For the system to work properly the chances of a stop pulse occurring should be small (<0.01–0.1/pulse). Under these conditions the probability of a stop pulse occurring at a specific time will be directly proportional to the probability of a photon being emitted by the sample at that time. Thus, if many flashes are analyzed, the contents of MCA channels will develop into a histogram having the shape of the sample decay curve. Figure 2-5 shows typical SPC transients.

The SPC method has numerous advantages. Since the probability of detecting a photon must be low, excitation intensity and optical efficiency are less important. SPC is, thus, one of the most sensitive approaches available. Unlike most other approaches, the system response is not limited to the minimum output pulse width of the PMT; the leading edge of the PMT pulse does the triggering, whereas the width and any tail are not used at all. The minimum pulse response is then controlled by jitter in the leading edge of the PMT pulse and in the trigger circuits and TAC. Thus, a nanosecond-wide pulse width is easily observed with a PMT having an impulse response several nanoseconds wide. The dynamic range and linearity of SPC is also enormous. As long as one is willing to wait to collect the counts, three or four orders of magnitude are common and five orders are possible.

Another major advantage of SPC is that the statistics are uniquely defined. SPC data exhibit nuclear counting statistics that obey a Poisson distribution (Bevington, 1969). A sample with a Poisson distribution exhibits a standard deviation equal to the mean sample count. For large counts the distribution goes over from an assymetrical Poisson one to an essentially perfect Guassian one. Thus, in data workup, the necessary statistical information is available for properly weighting each point. No other data collection method seems to have this advantage.

SPC has limitations, however. With available N_2- and D_2-filled flash lamps, excitation is most readily done below 400 nm. The H_2 and D_2 lamps can have subnanosecond pulse widths but achieve this only by supplying so few photons (10^8–10^9/flash) that even with the enormous sensitivity of SPC the intensity can be marginal. Further, flash lamps run at 10^4–10^5 Hz. If only 1% of the pulses are counted, the rate of collection is only 100–1000 counts/sec. To obtain tens of thousands of counts in 512 or 1024 channels then requires hours. With intense samples where fewer

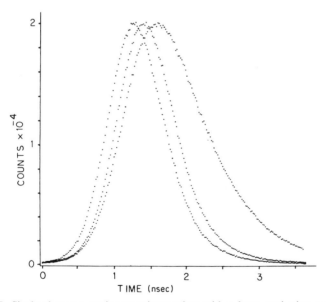

Fig. 2-5. Single-photon counting transients taken with a laser excitation source. The flash is the leftmost curve, and the two curves to the right are for rose bengal in water (τ = 114 psec) and methanol (τ = 543 psec). [Reprinted with permission from Cramer and Spears (1978). Copyright 1978 American Chemical Society.]

counts and/or fewer channels are required, acquisition is typically done in 5–15 min. Unlike many other data collection systems, the SPC instrument is largely limited to SPC. Finally, it is not as fast or as inexpensive as some other instruments.

Subnanosecond pulse width high-repetition-rate lasers and fast PMTs do permit direct studies on subnanosecond phenomena. But the cost and complexities of operating such systems greatly limit their general applicability.

Because of the relatively modest cost, high accuracy and precision, wide dynamic range, and excellent temporal resolution, SPC will continue to be the data acquisition method of choice of those requiring high resolution. Other methods such as oscilloscopes, sampling, and time shift will, for economic reasons and flexibility, still maintain a viable and useful position in solving many problems.

5. Phase Shift

The previously described techniques are all designed to record a transient following an excitation pulse. Quite different from these is the phase shift method. The sample is excited with a sinusoidally varying excitation. The sample decay will then vary sinusoidally, but the sample emission will generally lag behind the excitation waveform. The longer the sample lifetime, the more the emission is delayed and the greater the phase shift between the two waveforms. For a sample characterized by a single exponential decay time τ, the phase shift δ is given by

$$\delta = \tan^{-1}(2\pi f \tau) \tag{2-2}$$

where f is the modulation frequency (Chapter 4). Phase angles of high-frequency waveforms can be measured accurately, and extremely short lifetimes can be determined. Indeed, the first fluorescence decay times were measured by phase shift methods. Phase shift methods currently have temporal resolution in the low-picosecond range, which, at least for the moment, gives them a distinct advantage in speed over flash techniques.

A major disadvantage of the phase shift method is that it is difficult to use in analyzing complex kinetics. No matter how complex the kinetics, only a single phase shift is measured. This problem can be in part circumvented by making phase shift measurements at a number of frequencies, but the simplicity and ease of the phase shift measurement is then compromised. For further details of the phase shift method see Lakowicz (1983) and Spencer (1970).

C. FLASH PHOTOLYSIS

Flash photolysis is analogous to the ground state method of monitoring kinetic information by absorption spectroscopy. The difference is that one must establish a high excited state concentration by means of a short, intense pulse of irradiation. Flash photolysis has the advantage that the excited state does not have to emit, but not all excited molecules have strong enough excited state absorptions to be usable.

Figure 2-6 shows a typical low-speed flash photolysis system. The high excited state concentration is achieved by xenon flash lamps typically

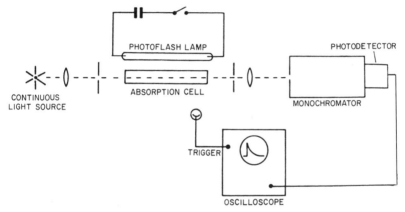

Fig. 2-6. Conventional flash photolysis instrument. [Reprinted from West (1976) by courtesy of Marcel Dekker, Inc.]

dissipating 50–10,000 J of energy in 2–200 μsec. The excited state concentrations are followed by a single-beam absorption instrument. A xenon flash system is suitable only for relatively long-lived transients, since the temporal resolution is limited by the long flash width.

Replacement of the flash lamp excitation by a pulsed laser with low-nanosecond pulse widths permits the measurement of nanosecond lifetime excited states. Because of the small laser beam areas, however, generally right angle monitoring of square cells or colinear monitoring along the laser beam are used (West, 1976).

D. PICOSECOND TECHNIQUES

Moving into the low-picosecond time domain requires complete modification of the excitation and detection approaches, although both absorption and emission monitoring can be used. The light sources are either mode-locked or synchronously pumped dye lasers. Picosecond techniques are generally much more expensive to implement and much more complex to maintain than the other approaches discussed.

If luminescence monitoring is used, even the fastest phototubes are inadequate and streak cameras must be employed. A streak camera is shown in Figure 2-7. A silt image of the incident light strikes a photocathode. The resultant photoelectrons are then accelerated into a deflection field. The beam is then swept linearly in time across a microchannel plate electron amplifier which (Csorba, 1980) amplifies the electron beam perhaps 1000 times while preserving the spatial information. This amplified electron beam strikes a phosphor screen which can be photographed or processed by a video camera–computer system. In the figure, the streak image develops down the phosphor screen. One is, in effect, converting temporal information into spatial information. Figure 2-8 shows a picosecond decay curve measured on a video camera-based streak camera system. The signal-to-noise ratio is good given the low-picosecond decay time. The temporal resolution of streak cameras is in the 2–100-psec range depending on the unit. Streak cameras are, however, quite expensive.

Absorption measurements are more difficult. Figure 2-9 shows a simple absorption picosecond instrument for monitoring absorption at wavelengths available from the excitation laser. In this case the 1060-nm picosecond pulse is frequency doubled and tripled. The 353-nm beam is used to excite the sample, and some of the 530-nm light is used to probe the excited state concentration. The two mirrors before and after the sample reflect part of the 530-nm beam onto a fast photodiode. The difference in the intensity of the two reflected beams provides a measure of the sample

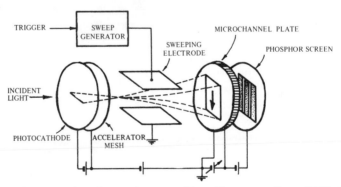

Fig. 2-7. Representation of a streak camera. [From Hamamatsu Corp. (1978). Courtesy of Hamamatsu TV Co.]

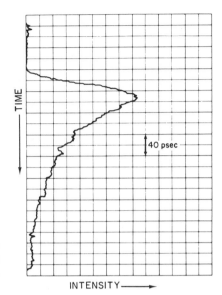

TIME

40 psec

INTENSITY ──▶

Fig. 2-8. Laser-excited picosecond decay of erythrosin in water captured with a streak camera. [From Hamamatsu Corp. (1978). Courtesy of Hamamatsu TV Co.]

absorbance at 530 nm. The two reflected beams can be monitored on the same detector, since they are separated in time. To obtain a complete decay curve one repeats the experiment a number of times while progressively delaying the arrival of the probe beam by moving the mirrors in the optical delay. One obtains an absorbance as a function of distance on the delay line, but from the speed of light (3.3 mm/psec) one obtains a decay versus time. Refinement of this basic technique permits the measuring of a number of delays in a single experiment (West, 1976).

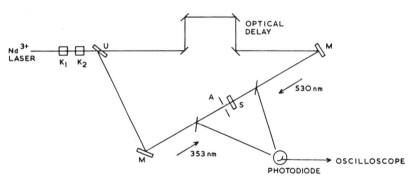

Fig. 2-9. Schematic representation of a picosecond decay time instrument using absorption monitoring of the induced transient. [Reprinted with permission from West (1976) by courtesy of Marcel Dekker, Inc.]

For extremely fast decays (<1–2 psec) an even more exotic approach is used. A subpicosecond pulse train from a synchronously pumped cw laser is split into a photolysis beam and a probe beam, as in Fig. 2-9, except that both are at the same wavelength. Now, instead of monitoring the probe beam before and after the sample, one monitors it only after the sample. To measure the absorbance one chops the photolysis beam at a low frequency and measures the modulated intensity of the unchopped probe beam with a lock-in amplifier. The resultant signal is the change in the percentage transmittance with the photolysis beam on and off. As in the previous example, absorbance versus time is obtained by mechanically delaying the arrival of the probe by increasing the distance. Figure 2-10 is the decay curve for the charge transfer excited states for a blue copper protein obtained with 0.5-psec excitation (Wiesenfeld *et al.*, 1980). There are two decay components. The fast one is instrumentally limited and is <~0.5 psec, whereas the slow (1.6 psec) one is easily resolved.

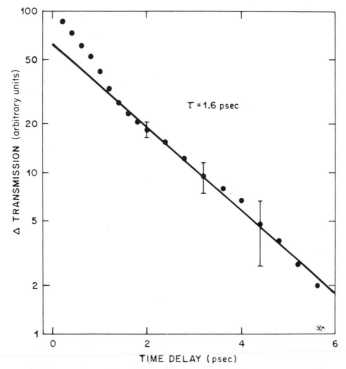

Fig. 2-10. Semilogarithmic plot of the molecular decay of the copper protein azurin. The fast initial decay is instrumental, whereas the slow step is the decay of the charge transfer excited state. [Reprinted with permission from Wiesenfeld *et al.* (1980). Copyright 1980 American Chemical Society.]

E. MISCELLANEOUS METHODS

For the measurement of long-lived phosphorescence, a rotating can phosphoroscope of the type used on spectrofluorimeters can be used. The phosphoroscope is generally a slotted can that spins at high speed around the sample. The slots are arranged so that, when excitation enters the sample the detector is blocked, and when the detector can see the sample the excitation is blocked. If the rotation speed is adjusted appropriately, the sample will still be emitting when the slot exposes the phototube to the sample, and the sample decay can be monitored. The speed must not be too fast, however, or the sample emission will not decay far enough during the phototube's open cycle.

An extremely interesting and potentially quite valuable way of studying lifetimes of nonluminescent species is photoacoustic spectroscopy (PAS). Chopped excitation is used, but instead of measuring the emission one follows an acoustic signal. Absorbed radiation is converted into heat which produces a pressure wave. If the excitation is chopped, the pressure wave is modulated and can be monitored with a microphone. The degree of modulation and its phase angle with respect to the excitation depend on the rates and pathways of relaxation much as in the phase shift luminescence method. The photoacoustic technique is particularly invaluable for studying radiationless processes, especially in gases. With standard equipment it should be possible to measure lifetimes in the range 10^{-5}–10^{-1} sec (Mandelis and Royce, 1980; Hunter and Stock, 1974).

Another limited approach uses emission polarization. Many excited molecules absorb and emit preferentially along a single molecular axis. If an ensemble of molecules is excited with polarized light, a preferential distribution of excited molecules results. If the rotational time is comparable to, or slower than, the lifetime, the emission will come preferentially from ordered molecules and will be polarized. From the degree of polarization and assumptions about the rotation characteristics, the lifetime can be estimated (Perrin, 1929).

Where the excited state cannot be monitored directly because it is nonluminescent or lacks absorptions in accessible regions, it is sometimes possible to measure the lifetime by monitoring the rate of consumption or formation of a product formed by reaction with this state. Absorption monitoring is usually used in a flash photolysis instrument. This approach has been used to measure the lifetimes of singlet oxygen (Merkel and Kearns, 1971; Young *et al.*, 1973) and stilbenes (Caldwell and Cao, 1981).

Simple Systems

A. INTRODUCTION

This chapter introduces a series of simple systems and concepts. The necessary mathematical tools come directly from elementary calculus, differential equations, and kinetics. Good references on kinetics include Laidler (1965), Benson (1960), Capellos and Bielski (1972), and Frost and Pearson (1953). Practical physical examples and inelegant mathematical approaches are used for clarity. Initially, as an aid to those who have been away from kinetics for a while, developments are given in greater detail. More elegant mathematical approaches are introduced in later chapters.

We begin by describing the more important (i.e., more common) possible fates of a species following excitation.

$$D + h\nu \xrightarrow{\ I\ } {}^*D, \tag{3-1a}$$

$${}^*D \xrightarrow{\ k_r\ } D + h\nu, \tag{3-1b}$$

$${}^*D \xrightarrow{\ k_q\ } D + \Delta, \tag{3-1c}$$

$${}^*D \xrightarrow{\ k_p\ } \text{products}, \tag{3-1d}$$

$$*D + Q \xrightarrow{k_{2a}} D + Q + \Delta, \qquad (3\text{-}1e)$$

$$\xrightarrow{k_{2b}} D^{\pm} + Q^{\mp}, \qquad (3\text{-}1f)$$

$$\xrightarrow{k_{2c}} D + *Q, \qquad (3\text{-}1g)$$

$$k_2 = k_{2a} + k_{2b} + k_{2c} \qquad (3\text{-}1h)$$

where D and *D are the ground and excited state form of the species under study. Q is a deactivator or quencher which can, on close interaction with the excited donor *D, deactivate or quench it by a catalytic step [Eq. (3-5e)], an excited state electron transfer [Eq. (3-5f)], or an energy transfer [Eq. (3-5g)]. $D^{\pm}$ and $Q^{\mp}$ are the oxidized and reduced forms of the donor and quencher, respectively. In general, the bimolecular processes may or may not involve diffusional motion of Q and D, depending on how the system is organized and the type of interactions. In this chapter we will consider only diffusional interactions. We have ignored interactions of one *D with another *D (excited state–excited state processes). Generally these can occur only under conditions of extremely intense illumination or with very long excited state lifetimes.

Electron transfer is a basic, critical excited state photochemical step in a variety of solar energy conversion systems and in photosynthesis (Clayton, 1971). For example, see the review by Grätzel (1981) on artificial photosynthesis. He discusses the half of the cycle where oxygen is generated and shows the variety of photochemical and electrochemical tricks needed to affect this seemingly simple reaction. For other examples of excited state electron transfer processes see the representative articles in the book by Hautala et al. (1979).

Excited state energy transfer is routinely used in synthetic photochemistry (Turro, 1978), in lasers (Schäfer, 1977), in solar energy conversion designs for increasing the efficiency of producing the active *Q, and in the molecular yardstick described in Chapter 1. A classic example of energy transfer is the production of singlet oxygen by energy transfer from dyes. Excited singlet oxygen is produced very efficiently on absorption of visible photons by a sensitizing dye in the presence of oxygen. The absorbing species can be an organic dye such as rose bengal or methylene blue (Schapp, 1976), or a metal complex (Demas et al., 1977b). This process is especially dramatic, since oxygen is colorless and cannot be directly excited except by light below 200 nm.

The rate or rate constants for each process are indicated above each equation: I is the rate of excited state production; we will use units of moles of excited state per liter per second (M sec^{-1}). Here k_r is the radiative rate constant for emission of a photon from the excited state; k_q is the rate constant for intramolecular excited state deactivation; k_p is the rate constant for direct excited state decomposition; and k_2 is the bimolecular rate constant for all bimolecular excited state deactivations. Within the model of Eqs. (3-1), the rate of production of excited state *D, $d[*D]/dt$, is given by

$$d[*D]/dt = I - k[*D] \qquad (3\text{-}2a)$$

$$k = k_r + k_q + k_p + k_2[Q] \qquad (3\text{-}2b)$$

k has units of a first-order rate constant (time^{-1}). Both I and [Q] may vary during the course of the experiment. In general, however, during an excited state measurement, [Q] remains virtually constant and then k can be treated as a constant. This situation occurs because the readily obtainable [*D]'s are generally much lower than [Q]. Under conditions where the bimolecular term is nonzero, but essentially constant, the system is referred to as pseudo–first order. In all subsequent discussions, k is assumed to be time independent.

B. FIRST-ORDER DECAYS

The majority of lifetime measurements involve first-order or pseudo-first-order kinetic processes. A first-order system is one that, in the absence of external perturbations such as the addition of more material, satisfies

$$d[*D]/dt = -k[*D] \qquad (3\text{-}3)$$

where Eq. (3-3) is written for an excited state. $d[*D]/dt$ has the same meaning as in Eq. (3-2), and k is a first-order rate constant (sec^{-1}) for loss of *D. For pseudo-first-order conditions with $I = 0$, Eq. (3-2) is of the form of Eq. (3-3). Consequently, any conclusions drawn about first-order processes apply to pseudo-first-order ones. k is no longer a pure first-order term, however, but now includes second-order components. First-order situations arise if the probability of decay of *D is unaffected by the presence of like neighbors. Normal radioactive decay is first order, since the probability of decay of each atom is unaffected by its neighbors. Many excited state relaxation processes satisfy Eq. (3-3), since [*D] is usually too low to affect the decay of *D.

If no *D is added to the system during the decay, the time dependence of [*D] in Eq. (3-3) is

$$\ln([*D]) = \ln([*D]_0) - kt \qquad (3\text{-}4a)$$

$$\log([*D]) = \log([*D]_0) - 2.303kt \qquad (3\text{-}4b)$$

$$[*D] = [*D]_0 \exp(-kt) \qquad (3\text{-}4c)$$

where $[*D]_0$ is the initial $[*D]$. The above system is approximated by a sample excited by a pulsed light source whose width is much shorter than the excited state lifetime. Formation of $*D$ is then complete before appreciable loss of $*D$ can occur. Equation (3-4) is the standard method of analyzing kinetic data. A $\ln([*D])$-versus-t plot is linear with a slope of $-k$ and an intercept of $\ln([*D]_0)$.

The methods of following $[*D]$ have been described briefly in Chapter 2. Flash photolysis methods with absorption monitoring of $[*D]$ are analogous to conventional, but slower, absorption methods of analysis of ground state reactions.

In luminescence monitoring of $[*D]$, the emission signal is frequently directly proportional to $[*D]$. This is shown as follows: An ideal detector's response is directly proportional to the rate that photons strike it, which in turn is proportional to the rate of photon emission:

$$\text{rate of emission of photons} = k_r[*D] \qquad (3\text{-}5)$$

Thus, $[*D]$ is proportional to the emission signal which can then frequently replace $[*D]$. Detector signals are not always proportional to $[*D]$. Many detectors are not ideal and can greatly distort the observed signal; see Chapter 7. Also, if the emission spectrum changes with time, then the detector's wavelength-dependent response invalidates the assumption by modulating the proportionality constant with time.

Figure 3-1 shows a typical luminescence decay curve and the linearized semilogarithmic plot. The decay is nearly exponential, but the small deviations are real and probably arise from different emitting sites in the crystals.

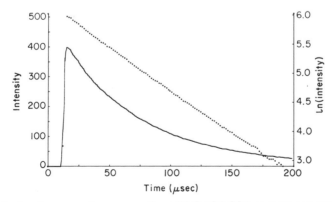

Fig. 3-1. Luminescence decay curve for solid $Cu(CF_3CO_2)$ at room temperature. The solid curve is the observed decay curve, and the dotted curve is the corresponding semilogarithmic plot. The lifetime in this case is approximately 65 μsec. The data were taken on a computer-controlled Tektronix 7912 transient digitizer. [Figure courtesy of Turley (1980).]

C. MEAN LIFE AND HALF-LIFE

The half-life $T_{1/2}$ is a pragmatic quantity defined as the time it takes the concentration to fall to half its original value. In excited state measurements this usually means the time required for the emission intensity to decrease by half. Although $T_{1/2}$ is easy to read from oscilloscope traces, it is of limited theoretical significance.

A more fundamental quantity is the mean lifetime τ which is sometimes also denoted by τ_m. For an exponential decay, we shall show that

$$\tau = 1/k \tag{3-6}$$

The initial [*D], [*D]$_0$, is reduced by a factor of $1/e$ at $t = \tau$. For an exponential decay, $T_{1/2}$ and τ are related by

$$T_{1/2} = (1/k) \ln 2 = \tau(\ln 2)\tau = 0.69319\tau \tag{3-7}$$

For an exponential decay, $T_{1/2}$ and τ are independent of the starting time t_0 where we choose [*D]$_0$. This is shown by comparing [*D] at any starting t_0 and at $t_0 + T_{1/2}$.

Equation (3-7) is widely used to estimate τ's from oscilloscope or recorder traces. It is easy to pick a point on a trace and note the time required for the signal to decay to half this initial value, which yields $T_{1/2}$ and then τ from Eq. (3-7). Since for an exponential decay $T_{1/2}$ is independent of t_0, the invariance of $T_{1/2}$ with t_0 is a simple way of estimating whether a decay is exponential without having to resort to semilogarithmic plots. We demonstrate the usefulness of this technique in Section G.

We now prove Eq. (3-6) relating τ and k for an exponential decay. The mean lifetime τ is just the statistical average lifetime as used in population analysis. Thus, τ is the summation over all time of the number of members lost in a time interval Δt times their age at loss; this quantity is averaged over the total population. In integral form

$$\tau = \int_0^\infty t(-d[*D]/dt)\ dt \Big/ \int_0^\infty (-d[*D]/dt)\ dt \tag{3-8}$$

For a first-order decay $-d[*D]/dt$ is given by Eq. (3-3) and [*D] by Eq. (3-4c) which on substitution into, and integration of, Eq. (3-8) yields Eq. (3-6). Here τ is larger than $T_{1/2}$, since an exponential tails out to infinity and the molecules decaying at long times weight τ toward longer times.

Equation (3-6) applies only to an exponential decay. For nonexponential decays τ is still defined by Eq. (3-8), but integration is usually carried out numerically. For nonexponential decays, τ's calculated from Eq. (3-8) are more meaningful than $T_{1/2}$'s, since a small component of a fast decay produces a short $T_{1/2}$ even though τ is quite long.

D. STEP EXCITATION

We now consider a more complex system. What is the response of an initially unexcited sample subjected to a constant-intensity excitation? We turn the excitation on at $t = 0$ with a constant rate of generating *D of I (M sec^{-1}). Again, the system is first or pseudo–first order. This configuration is the common step excitation method of decay time measurements: Open the shutter between a continuous arc lamp and a sample, let the sample glow brightly, close the shutter, and record the decay curve. An important question is, How long should the shutter be held open? The following analysis shows that it should be held open for at least 2τ to 3τ to permit the sample to approach its terminal emission intensity.

The differential equation for [*D] is Eq. (3-2) with a constant I. The solution with [*D] $= 0$ at $t = 0$ is

$$\int_0^{[*D]} \{1/(I/k - [*D])\}\, d[*D] = k \int_0^t dt \tag{3-9a}$$

$$[*D] = [D]_\infty[1 - \exp(-kt)] \tag{3-9b}$$

$$[D]_\infty = I/k \tag{3-9c}$$

At very large times ($kt \gg 1$), [*D] levels off at $[D]_\infty$. This saturation occurs because the rate of production of [*D] is constant but the loss is proportional to [*D]. Eventually, [*D] becomes large enough so that the losses equal the gains, and a steady state or constant value of [*D] results. When [*D] is monitored by luminescence with an ideal detector, the observed decay curve $D(t)$ is

$$D(t) = D_\infty[1 - \exp(-kt)] \tag{3-10a}$$

$$\ln[1 - (D(t)/D_\infty)] = -kt \tag{3-10b}$$

D_∞ is the limiting $D(t)$ at $t = \infty$. Equation (3-10b) is the linearized form used for evaluating k. A $\ln(1 - D(t)/D_\infty)$-versus-t plot will be linear with a slope of $-k$ and an intercept of 0 if the system is first order. Also, $\tau = 1/k$. For $D(t)$ to achieve a substantial fraction of its terminal value, the time required is at least 2τ to 3τ.

Equation (3-10) can fail if τ is very long and/or the excitation source is very intense. If a large fraction of D is excited, then I can fall either because the decrease in [D] reduces the absorbance and lets excitation light escape or because *D can compete with D for the available photons. Also, with the strongest excited state pumping, population inversion and stimulated emission can arise. This last effect is responsible for the laser and can cause temporal and positional oscillations in [*D].

E. QUENCHING AND LUMINESCENCE QUANTUM YIELDS

So far we have not considered the effect of nonradiative pathways [Eqs. (3-1c) to (3-1g)] on the system's luminescence. We now show that the nonradiative pathways reduce the probability of emission of a photon and develop an important relationship between τ, the emission probability, and the radiative rate constant k_r.

All our previous conclusions are true for first and pseudo-first-order processes. For example, decays will be exponential regardless of k_q or k_2. Further, even though we indicated only one internal first-order quenching pathway, the inclusion of any number of them does not affect our previous conclusions. For example, assume there are a number of first-order quenching pathways k_{qi}, then Eq. (3-2) ($I = 0$) becomes

$$d[*D]/dt = -(k_r + k_p + k_2[Q] + \Sigma k_{qi})[*D] \qquad (3\text{-}11)$$

We can define k_q as

$$k_q = \Sigma k_{qi} \qquad (3\text{-}12)$$

where the summation is over all the k_{qi}. These k_{qi} could represent a number of radiationless pathways for decay from a manifold of sublevels in thermal equilibrium. Thus, by monitoring a decay curve, no direct information is obtained about the relative magnitudes of k_r, k_p, k_2, and k_{qi}, or even the number of k_{qi}.

How does one obtain the fundamentally important k_r and k_q? In addition to τ, the luminescence photon or quantum yield is required. The photon yield Q is the probability that an excited molecule will emit. Q can range from zero for nonluminescent species to near unity for fluorescein, perylene, and a few laser dyes. Q equals the ratio of the rate of photon emission $k_r[*D]$ to the total rate of excited state loss ($k_r + k_q + k_p + k_2[Q])[*D]$. These two rates are independent of the time dependence of I, and

$$Q = k_r/(k_r + k_q + k_p + k_2[Q]) \qquad (3\text{-}13a)$$

$$Q = k_r\tau \qquad (3\text{-}13b)$$

where the second form results from Eq. (3-6). The radiative or intrinsic lifetime τ_r is given by

$$\tau_r = 1/k_r \tag{3-14}$$

τ_r is the lifetime that would be observed if no quenching or photochemistry were present and is sometimes called the natural lifetime. We avoid this usage, since to some it incorrectly implies an observed quantity. Also, τ_0 is frequently used in place of τ_r, but τ_0 has another widely accepted use (Section F). In the absence of photochemistry and bimolecular quenching, k_q is also readily obtained from τ and Q:

$$k_q = (1 - Q)/\tau \tag{3-15}$$

which is Eq. (3-13) rearranged.

The fundamental, practical, and theoretical importance of k_r and k_q explains, in part, the amount of work involving Q and τ determinations. Numerous elaborate and ingenious experimental methods are available for obtaining Q's. This complex problem is reviewed elsewhere (Demas and Crosby, 1971; Crosby et al., 1972; Demas, 1982).

In the past, good experimental Q's were difficult to measure. Equation (3-13b) was sometimes used for estimating Q's. Theoretical treatments are available for estimating k_r from the sample's absorption and/or emission spectra (Strickler and Berg, 1962). Thus k_r, τ, and Eq. (3-13b) permitted estimation of Q. For fluorescences derived from a single highly allowed transition, this procedure gives reasonable, sometimes even remarkably good, Q's.

This approach fell into error when it was used to disprove published Q's. With few exceptions the use of such an approximate theoretical treatment to disprove well-designed and well-executed experiments is unsound. We stress "well-designed" because of the numerous abominably bad Q determinations in the literature. Fortunately, instrumentation and techniques for measuring Q's are steadily improving, and it is becoming increasingly harder to publish and defend claims based solely on calculated Q's.

F. BIMOLECULAR QUENCHING AND THE STERN–VOLMER LAW

We now turn to bimolecular quenching [Eqs. (3-1e) to (3-1g)] and the Stern–Volmer (SV) formula (Stern and Volmer, 1919; Turro, 1978). We explicitly consider the second-order terms in Eqs. (3-1e) to (3-1g) and continue to assume that they are pseudo–first order.

Many substances efficiently quench excited states by a bimolecular collisional process. For example, oxygen is a notoriously good quencher of the fluorescence and phosphorescence of many organic molecules (Birks, 1970), as well as the emissions of inorganic ones (Demas et al., 1977b). As a result, photochemical and photophysical work done in fluid solutions frequently employs elaborate deoxygenation methods such as multiple freeze–pump–thaw cycles on a vacuum line.

There is great interest in both energy and electron transfer quenching processes, and quantitative k_2 are essential for their study. Bimolecular rate constants can readily be obtained from the relative luminescence yield or lifetime of a sample versus the quencher concentration (Stern–Volmer equations).

For demonstration purposes we develop the Stern–Volmer equation using a steady state analysis. Under conditions of steady illumination [*D] in Eq. (3-2) quickly becomes stable if k_p is small compared to the other terms (e.g., $d[*D]/dt = 0$). Solving for [*D] yields

$$[*D] = I/(k_r + k_p + k_q + k_2[Q]) \tag{3-16}$$

Since the observed emission intensity ϕ is proportional to $[*D]k_r$,

$$\phi = I'/\{k_r + k_p + k_q + k_2[Q]\} \tag{3-17}$$

where I' includes k_r and the necessary proportionality constant for converting to the measured units and conditions. Generally, ϕ is obtained by monitoring the sample emission at a single-emission wavelength.

Equation (3-17) is converted to a form useful for data treatment by eliminating I'. Typically, a series of measurements of ϕ's versus [Q] are made. The first one, with [Q] = 0, yields an initial intensity ϕ_0.

$$\phi_0 = I'/(k_r + k_p + k_q) \tag{3-18}$$

Combining Eqs. (3-17) and (3-18) yields

$$(\phi_0/\phi) - 1 = K_{SV}[Q] \tag{3-19a}$$

$$K_{SV} = k_2/(k_r + k_p + k_q) = k_2\tau_0 \tag{3-19b}$$

where K_{SV} is the Stern–Volmer quenching constant and has units of reciprocal concentration, τ_0 is the unquenched lifetime at [Q] = 0, and K_{SV} is determined from the slope of the linear plot of $(\phi_0/\phi) - 1$ versus [Q]. The larger K_{SV}, the more efficient the quencher. K_{SV}^{-1} equals the quencher concentration necessary to produce a 50% decrease in the emission intensity.

K_{SV} provides a practical measure of quenching efficiency and is generally easy to measure, but it lacks insight into the fundamentally important

k_2. For example, a species with a long τ_0 can yield a large K_{SV} even if k_2 is relatively small, whereas a very high k_2 will be masked by a short τ_0. Values of k_2 can be calculated from Eq. (3-19b) if τ_0 is also measured.

How safe was the steady state assumption used in Eq. (3-16)? Asked in another manner, How soon after placing the sample in the spectrofluorimeter is $d[*D]/dt = 0$? Usually far faster than the data are taken; therefore, the approximation is invariably good. For purposes of analysis we assume a step excitation. From Eq. (3-9), [*D] rises exponentially to the steady state value with a lifetime equal to the sample lifetime. Thus, even if $\tau = 100$ msec, which is an exceptionally long lifetime in fluid solutions, the intensity will be within 1% of the steady state value in 500 msec (5τ) and within 0.01% in 1 sec (10τ).

K_{SV} and k_2 can also be measured from τ-versus-[Q] data. Combining Eqs. (3-2c) and (3-6) gives

$$1/\tau = (k_r + k_p + k_q) + k_2[Q] \qquad (3\text{-}20a)$$

$$1/\tau = (1/\tau_0) + k_2[Q] \qquad (3\text{-}20b)$$

$$(\tau_0/\tau) - 1 = k_2\tau_0[Q] = K_{SV}[Q] \qquad (3\text{-}20c)$$

Equations (3-20) are the lifetime forms of the Stern–Volmer equation and can supply the same information as the intensity form Eq. (3-19). Indeed, plots of $(\tau_0/\tau) - 1$ and $(\phi_0/\phi) - 1$ versus [Q] should be indistinguishable if the simple diffusional quenching model holds.

Figures 3-2 to 3-4 are typical Stern–Volmer plots using the intensity and lifetime quenching methods. The luminescent species is the now widely used inorganic complex [Ru(bpy)$_3$]$^{2+}$ (bpy = 2,2'-bipyridine) which has $\tau_0 = 600$ nsec. The quenchers are PtCl$_4^{2-}$, HgCl$_2$, and O$_2$. The mechanism of quenching for PtCl$_4^{2-}$ (Demas and Adamson, 1971) is simple energy transfer [Eq. (3-1g)], whereas HgCl$_2$ quenches by electron transfer [Eq. (3-1f)] from the Ru(II) species to HgCl$_2$ (DeGraff and Demas, 1980; Hauenstein et al., 1983b). O$_2$ quenching yields singlet O$_2$ with a very high efficiency (~65–85%), but there is some question whether an energy (Demas et al., 1977b) or an electron transfer process (Lin and Sutin, 1976) is involved, although this author favors energy transfer. The O$_2$ quenching results use the τ Stern–Volmer equation and were obtained with a $150 submicrosecond decay time system used in a physical chemistry laboratory experiment (Demas, 1976). The ϕ and τ quenching data agree within experimental error. As shown later, discrepancies between the ϕ and τ Stern–Volmer plots can supply useful chemical information.

Equation (3-20b) can be used to evaluate k_2 directly; see Fig. 3-4. Until the development of computerized data acquisition systems with on-line data reduction capabilities, this approach was generally unattractive be-

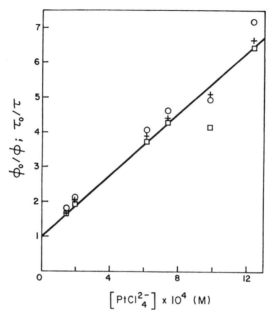

Fig. 3-2. Quenching of the emission of Ru(bipy)$_3$Cl$_2$ in deoxygenated 1×10^{-3} M HClO$_4$ by PtCl$_4^{2-}$. +, Lifetime data ($\tau_0 = 0.685$ usec); $\bigcirc$ and $\square$, relative intensity data taken before and after the lifetime measurements, respectively. [Reprinted with permission from Demas and Adamson (1971). Copyright 1972 American Chemical Society.]

cause of heavy computational demands. There are, however, times when only τ data can be used conveniently. These include situations in which the quencher absorbs intensely in the region of excitation and/or emission. Under these conditions, trivial absorption can produce unacceptable distortions of intensity quenching data unless complicated, and frequently unreliable, corrections are applied (Demas and Adamson, 1973; Leese and Wehry, 1978). Lifetime measurements are free of these artifacts, since the shape of the decay curve is unchanged.

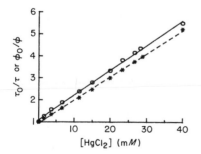

Fig. 3-3. Luminescence quenching of Ru(bipy)$_3$Cl$_2$ by HgCl$_2$ using lifetime ($\bigcirc$) and intensity (∗) data. The measurements are at room temperature. [Figure courtesy of B. L. Hauenstein, Jr. and K. Mandal.]

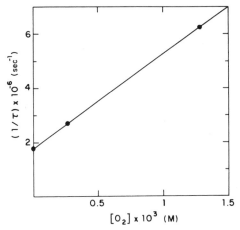

Fig. 3-4. Decay time Stern–Volmer plot for quenching of Ru(bipy)$_3^{2+}$ by oxygen. Results were taken on a $150 submicrosecond decay time instrument. [Reprinted with permission from Demas (1976). Copyright 1976 by the Division of Chemical Education, American Chemical Society.]

G. MULTICOMPONENT DECAYS

A common kinetic problem arises when several species with different lifetimes emit simultaneously and independently. Such systems usually arise from impure samples or mixtures, although pure systems can also exhibit such complex kinetics; see Chapter 4. For a delta function (i.e., zero duration) excitation with independently emitting noninteracting species, the detector response is

$$D(t) = \sum_{i=1}^{N} K_i \exp(-t/\tau_i) \qquad (3\text{-}21a)$$

$$\tau_i = 1/k_i \qquad (3\text{-}21b)$$

where N is the number of emitting components. For the ith component, τ_i is the lifetime and K_i is the preexponential factor contributing to the signal at zero time. The K_i are functions of the spectral response of the detector, the concentration, emission, and absorption characteristics of each component, the spectral transmission properties of the filters, and the spectral distribution of the exciting light. It is also assumed that the emission spectrum of each component is independent of time following the excitation.

In general, semilogarithmic plots of $D(t)$ versus t are concave upward for a multicomponent system unless the τ_i are nearly equal or one of the K_i dominates. The plots are then nearly linear.

Figure 3-5 shows $D(t)$ and $\log[D(t)]$-versus-t plots for a decay that is a sum of two exponentials ($K_1 = 5000$; $\tau_1 = 35$ nsec; $K_2 = 5000$; $\tau_2 = 7$ nsec). Visually, the $D(t)$ plot looks exponential. That it is not exponential is readily seen by applying the $T_{1/2}$ test to different portions of the curve. When $D(t)$ at $t = 0$ is used as the initial intensity, $T_{1/2} = 10$ nsec. Taking the initial intensity at $t = 10$ nsec yields $T_{1/2} = 14$ nsec. With the initial value at 50 nsec, $T_{1/2} = 25$ nsec. This last value corresponds to $\tau = 36$ nsec which is in excellent agreement with the long-lived component. This result is expected, as the 7-nsec component makes no significant contribution at these long times. The large discrepancies among the $T_{1/2}$ show that the decay is nonexponential. The nonlinear $\log[D(t)]$-versus-t curve (Fig. 3-5b) confirms this.

If only two components are present and their lifetimes differ significantly, it is sometimes possible to obtain the τ_i and K_i by component stripping just as is done in radiochemistry (Friedlander *et al.*, 1964). This procedure is as follows: Without loss of generality we denote the long-lived component as component 1. A $\ln[D(t)]$-versus-t plot is made. At long enough times this curve will be linear with a slope of $-1/\tau_1$ and an intercept of $\ln K_1$. Using this K_1 and τ_1, the contribution of the slow component is subtracted from $D(t)$ to yield $D'(t)$:

$$D'(t) = D(t) - K_1 \exp(-t/\tau_1) \qquad (3\text{-}22)$$

$D'(t)$ approximates the pure decay curve for the short-lived component and is reduced just as any normal single exponential decay curve. This procedure is demonstrated in Fig. 3-5b. Note that $D'(t)$ is not the difference between the extrapolated and observed semilogarithmic plots of Fig.

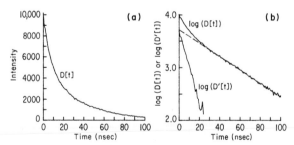

Fig. 3-5. A demonstration of component stripping in the evaluation of the sum of two exponentials. (a) Experimental decay curve $D(t)$. (b) Semilogarithmic plot of the data of (a), the least squares fit for the latter portion of this curve (---), and the semilogarithmic plot of the long-lived component-corrected decay $D'(t)$.

3-5b, but the difference between the observed decay and the calculated exponential decay of the long-lived component.

For the data of Fig. 3-5, the long-lived component was evaluated by fitting an unweighted linear least squares line (Chapter 5) to the $\log[D(t)]$-versus-t plot over the 50–99-nsec range to yield $K_1 = 5230$ and $\tau_1 = 34.2$ nsec. This fit and its contribution at earlier times are shown by the dotted line in Fig. 3-5b. Then $D'(t)$ was evaluated using Eq. (3-22). A least squares fit was made to the linear $\log[D'(t)]$-versus-t plot (Fig. 3-5b), but the fitting region was limited to 0 and 15 nsec because of the high noise levels at longer times. For this short-lived component $K_2 = 4774$, $\tau_2 = 6.71$ nsec. All four calculated parameters are in reasonable agreement with the values used to generate the curve.

Inspection of these plots reveals a danger in the component-stripping procedure. The $\log[D'(t)]$ plot versus time becomes excruciatingly noisy after about 20 nsec. Clearly, if much of the data beyond 20 nsec were used in the $D'(t)$ fit, poor results would be obtained. Indeed, beyond ~28 nsec, $D'(t)$ becomes negative, since $D(t)$ has virtually no information about the short-lived component in this region. Therefore, the fitting regions must be selected carefully.

To demonstrate in greater detail the utility and limitations of the component-stripping method, we have carried out fits on a number of data sets

TABLE 3-1

Results of Double-Exponential Fitting by Component Stripping

Actual values[a]		Calculated values[b]			
K_1	K_2	K_1	τ_1	K_2	τ_2
500	9500	564 ± 55	33.3 ± 1.6	9417 ± 89	7.0 ± 0.1
1000	9000	1053 ± 81	34.2 ± 1.3	8943 ± 80	7.0 ± 0.1
2000	1000	1979 ± 82	35.2 ± 0.7	7983 ± 77	7.1 ± 0.1
5000	5000	5024 ± 118	34.9 ± 0.4	5000 ± 115	6.9 ± 0.3
7000	3000	7020 ± 194	35.0 ± 0.5	3043 ± 157	6.7 ± 0.9
8000	2000	7974 ± 178	35.1 ± 0.4	2032 ± 166	6.9 ± 1.0
9000[c]	1000[c]	8994 ± 192	35.0 ± 0.4	1095 ± 167	6.5 ± 2.5
5000[d]	5000[d]	6279 ± 159	32.5 ± 0.4	3752 ± 162	12.8 ± 0.5

[a] Unless otherwise indicated, $\tau_1 = 35$ nsec and $\tau_2 = 7$ nsec, with the fitting region for the long component being 49–100 nsec and for the short component 0–14 nsec.

[b] Each entry is the mean and standard deviation for 20 simulations. Noise was added to synthetic noise-free decay curves to simulate single-photon counting statistics.

[c] For the fast component fit the fitting region was reduced to 0–9 nsec. Otherwise negative $D'(t)$'s resulted.

[d] $\tau_1 = 35.0$ nsec and $\tau_2 = 15$ nsec.

that are the sum of two exponentials with differing K's and τ's. In each case 20 different simulated data sets were used. These data simulated single-photon counting experiments with $K_1 + K_2 = 10,000$ counts (Chapter 11). Unweighted linear least squares fits were made to the appropriate regions of the $\ln[D(t)]$- or $\ln[D'(t)]$-versus-t plots. Table 3-1 summarizes the results; a mean and standard deviation were calculated for every parameter using the results of all 20 simulations. The component-stripping method works best if the τ's are well separated and the disparity between the K's is not large. As the τ's approach each other, accurate resolution, even with reasonably favorable K's, becomes increasingly difficult.

While component stripping has been used beyond two components, it becomes increasingly more difficult and places more stringent demands on the separation of the τ_i's and on the data quality. Generally, for more than two components, nonlinear least squares methods (Chapter 5) are used, but it is still a difficult problem. Also, as we shall show, nonlinear least squares is more accurate and precise than the component-stripping method for two components, but it is computationally more difficult.

CHAPTER

4

More Complex Systems

A. INTRODUCTION

In this chapter we treat more complex systems including several that have counterparts in common chemical kinetic systems (e.g., serial decay), as well as others that do not. The latter include the arbitrary excitation pulse, coupled systems with feedback (e.g., excimers and excited state acid–base reactions), and Förster long-range energy transfer. Space prevents us from discussing the complex but increasingly important areas of micelles, monolayers, and bilayers. The interested reader is referred to the review by Grätzel and Thomas (1976) and recent examples from the chemical literature (Kalyanasundaram, 1978; Almgren et al., 1979; Atik and Thomas, 1981).

B. ARBITRARY EXCITATION FUNCTIONS

So far we have assumed a delta function flash. We now consider a more realistic situation where the excitation source is of finite duration and may be comparable to or longer than the sample τ. For example, this problem arises when microsecond pulse duration xenon flash lamps are used to measure luminescence lifetimes of short-lived metal complexes. In the much faster time domain, the measurement of 0.1–20-nsec fluorescence lifetimes using 1–10-nsec-wide excitation pulses routinely presents exactly the same problem.

In this section we derive the luminescence response of a system that decays by first-order and/or pseudo-first-order processes when excited by an arbitrary excitation pulse. First, we confront the problem of predicting a decay curve given an excitation profile and a lifetime. Since, in general, we do not know the sample lifetime, we then explore the practical problem of extracting a lifetime from the observed flash and decay curves. We describe briefly how this process, called deconvolution, can be carried out if the sample is characterized by a single decay time. More detailed analyses of deconvolution methods, as well as treatment of more complex decay schemes, is deferred until Chapters 8 and 10.

If the sample's impulse response (the response to a delta function or zero duration pulse) is a simple exponential, $\exp(-kt)$, how does the observed decay $D(t)$ depend on τ and any arbitrary flash function $E(t)$? Usually $D(t)$ cannot be approximated by an analytical function, and numerical solutions become necessary. These computations are too complex for hand calculations, but the wide availability of low-cost, powerful programmable calculators and mini- and microcomputers has all but eliminated this bothersome computational problem.

If *D is the emitting species, then we begin with the differential equation for *D [Eq. (3-3)]:

$$d[*D]/dt = I(t) - k[*D] \qquad (4\text{-}1)$$

where [*D] is the excited state concentration at time t and k is given by Eq. (3-3c). Now $I(t)$, the rate of formation of *D is explicitly considered to be time dependent. Equation (4-1) is a linear first-order differential equation having the following exact solution (Appendix A):

$$[*D] = e^{-kt} \int_{x=0}^{x=t} \exp(kx)I(x)\, dx \qquad (4\text{-}2)$$

where x is a dummy variable of integration. It is common to measure the excitation profile and the sample decay with the same detector. Assuming an ideal detector, $D(t)$ is proportional to [*D] and $E(t)$ to $I(t)$, which yields

$$D(t) = Ke^{-kt} \int_{x=0}^{x=t} \exp(kx)E(x)\, dx \qquad (4\text{-}3a)$$

$$k = 1/\tau \qquad (4\text{-}3b)$$

where K is the proportionality constant matching the two curves.

Equation (4-3) makes several important assumptions. First, there is no time shift between $E(t)$ and $D(t)$. The use of internal triggering on an oscilloscope to measure $E(t)$ and $D(t)$ invalidates this assumption (Chapter 10). The excitation source must not be so intense that the effective viewing geometry and thus the proportionality constant relating $D(t)$ and [*D] change with time. Finally, although we have assumed an ideal detector, we will show in Chapter 7 that Eq. (4-3) is correct for a distorting detector as long as the distortions are "linear" and $E(t)$ and $D(t)$ are measured under the same conditions of distortion.

Figure 4-1 shows a typical $E(t)$ and the expected responses for samples of different lifetimes. The $E(t)$ selected has characteristics similar to those of a typical single-photon counting flash and exhibits a pulse width of 1.4 nsec (FWHM). The display curves are noise-free. Clearly for sample lifetimes longer than ~2.5 nsec, $D(t)$ is only slightly distorted by $E(t)$. For shorter lifetimes, however, the continued pumping by $E(t)$ has a profound effect on the $D(t)$'s, and the traditional method of making semilogarithmic plots of $D(t)$ versus t will not yield the lifetime of the decay. It is noteworthy that, even for sample lifetimes as short as 0.1 nsec, which is almost ~15 times shorter than the excitation pulse width, $D(t)$ is still appreciably different from $E(t)$ and thus carries information about the sample decay time. Indeed, a τ of 0.05 nsec yields a $D(t)$ that is appreciably different from $E(t)$, and even a 0.01-nsec lifetime gives a $D(t)$ that is discernibly different from $E(t)$.

Equation (4-3) is commonly used to estimate sample τ's where there is a discernible difference between $D(t)$ and $E(t)$ but τ is too short to use the semilogarithmic method. A simple procedure is to guess a series of likely τ's, calculate the expected decay curve $D^{calc}(t)$ from Eq. (4-3), and compare this with the observed $D(t)$. Then K is picked to give a good fit; frequently $D^{calc}(t)$ and $D(t)$ are scaled to either equal heights or, more accurately, equal areas. The process of guessing and comparing is repeated until the experimenter tires, the power goes off, an acceptable fit is arrived at, or it is concluded that the data cannot be fit by Eq. (4-3). The τ

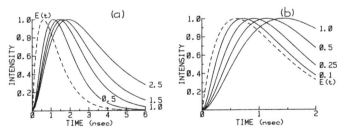

Fig. 4-1. Excitation flash $E(t)$ (---) and observed decay curves $D(t)$ (——) for luminescent samples with different lifetimes. $E(t)$ is the same for (a) and (b), but a shorter time period is used for (b). The lifetime used to calculate each decay curve is given by the curve. Even for $\tau = 0.05$ nsec, the decay is easily discernible from $E(t)$.

yielding the best fit is decreed to be the sample lifetime. Generally the best fit is judged either visually or from a goodness-of-fit parameter that is the sum of the squares of the differences between $D^{calc}(t)$ and $D(t)$. This basic technique has several disadvantages. First, if a visual fitting is used, the results are dependent upon the observer's prejudice. Second, a problem common to all deconvolutions is that $E(t)$ should be reproducible from flash to flash since it is seldom possible to obtain $E(t)$ and $D(t)$ simultaneously. Finally, the number of integrals to be evaluated is rather formidable except when a computer is used, since each value of $D(t)$ and each trial K requires the evaluation of a numerical integral. We will return to this exceptionally important process of extracting kinetic information from $E(t)$ and $D(t)$ in Chapter 8.

It should be noted that Eq. (4-3) is perfectly general for an exponential decay function and that earlier results can be derived directly from it. For example, if $E(t)$ is zero before time zero and some constant value from $0 \leq t$ to infinity, Eq. (3-11) for a simple step function excitation immediately results. If $E(t)$ is a narrow spike at $t = 0$, then the integral is constant for $t > 0$ and a simple exponential decay results.

C. SUCCESSIVE FIRST-ORDER DECAYS

We first consider the commonly occurring case of successive first-order decays or series decays in which one excited species decays by a first-order or pseudo-first-order process to another excited species which in turn decays to a third species, etc.

$$A \xrightarrow{k_{AB}} B \qquad (4\text{-}4a)$$

$$B \xrightarrow{k_{BC}} C \qquad (4\text{-}4b)$$

$$C \xrightarrow{k_{CD}} D, \text{ etc.} \qquad (4\text{-}4c)$$

where A, B, C, and D represent different excited state species or, in the last step, the terminal ground state form. The k's, which are usually all different, represent the first-order or pseudo-first-order rate constant for each relaxation and are given by terms similar to those in Eq. (3-3c).

We point out that the nomenclature used for the rate constants of Eqs. (4-4) is the opposite of that used by Birks (1970). Birks and others following his lead employ an inverted notation where k_{BA} is used in Eq. (4-4a). We have adopted the above convention since it seems both more logical and is the one favored by virtually all kinetics treatments.

We now consider the simplest and most common case where the chain ends at C—typically because C is the ground state. We assume that after $t = 0$ no additional A, B, or C is added from external sources. This condition corresponds to pulse excitation with a very fast pulse. The relevant equations are

$$d[A]/dt = -k_{AB}[A] \tag{4-5a}$$

$$d[B]/dt = k_{AB}[A] - k_{BC}[B] \tag{4-5b}$$

Equation (4-5a) is a simple first-order decay and yields

$$[A] = [A]_0 \exp(-k_{AB}t) \tag{4-6}$$

where $[A]_0$ is the concentration of A at $t = 0$. Here [B] follows no simple or obvious expression but is a function of $[A]_0$, $[B]_0$, and the k's. Equation (4-5b) is a simple, linear, first-order differential equation that yields (Capellos and Bielski, 1972; Frost and Pearson, 1953)

$$[B] = P[\exp(-k_{AB}t) - \exp(-k_{BC}t)] + [B]_0 \exp(-k_{BC}t)$$
$$P = k_{AB}[A]_0/(k_{BC} - k_{AB}), \quad \tau_A = 1/k_{AB}, \quad \tau_B = 1/k_{BC} \tag{4-7}$$

where $[A]_0$ and $[B]_0$ have the same meaning as before and τ_A and τ_B are the mean lifetimes for the decay of A and B, respectively. Equations (4-7) can also be derived from one limiting case of the general equation for coupled systems (Section G).

If $[B]_0 = 0$, then [B] starts at zero, rises to a maximum, and eventually decays to zero. Figure 4-2 shows a series of decay curves for [A] and [B]. Here τ_A is taken as 20 nsec and τ_B is varied. Note that the maximum in B occurs at later and later times as τ_B/τ_A increases. The time required for the occurrence of this maximum is readily calculated. For a maximum $d[B]/dt = 0$. The time t_{max} satisfying this condition is

$$t_{max} = [\ln(k_{BC}/k_{AB})]/(k_{BC} - k_{AB}) \tag{4-8a}$$

$$= \tau_A\tau_B[\ln(\tau_A/\tau_B)]/(\tau_A - \tau_B) \tag{4-8b}$$

Figure 4-3 shows a plot of t_{max} versus τ_B/τ_A in units of τ_A. As τ_B/τ_A becomes larger, t_{max} moves to longer times. In the limit of $\tau_B/\tau_A = \infty$, the peak is never reached, which is exactly what one expects, since B is just accumulating (integrating) all the decay products of A, which in turn takes forever to decay completely. Note that t_{max} is most sensitive to variations in τ_B for small τ_B/τ_A. As τ_B increases beyond τ_A, however, progressively less information is contained in τ_{max}.

Equation (4-8) has been used by Kleinerman and Choi (1968) to measure rates of energy transfer from Tb^{3+} complex hosts to Eu^{3+} complexes

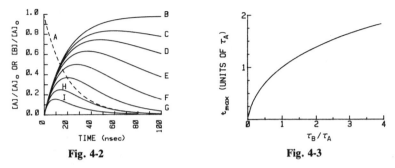

Fig. 4-2   **Fig. 4-3**

Fig. 4-2. Decay curves for the system A → B → C. In all cases $k_{AB} = 5 \times 10^7$ sec^{-1}. A (−−−) is the decay curve for [A]. Curves B–I are the decays of B for different values of k_{AB}/k_{BC}. The curves correspond to the indicated ratios: B, 200/1; C, 16/1; D, 8/1; E, 4/1; F, 2/1; G, 1.01/1; H, 1/2; I, 1/4. Curve B is indistinguishable from the case of $k_{BC} = 0$.

Fig. 4-3. Time of the maximum concentration t_{max} of B in an A → B → C system. τ_A and τ_B are the lifetimes of A and B, respectively, and t_{max} is in units of τ_A.

doped into the crystalline host. Figure 4-4 shows a typical decay curve. The slow rise and peaking is characteristic of Eq. (4-7). The long tail can be used to evaluate τ_B which corresponds to the lifetime of the Eu^{3+} guest. From t_{max} and Eq. (4-8b), τ_A (the Tb^{3+} lifetime) can be calculated. Comparison of this τ_A with that measured directly for the Tb^{3+} host shows that most of the Tb^{3+} complexes are capable of transferring energy to Eu^{3+}. This elegant method is regrettably of rather limited usefulness. Generally $k_{AB} \gg k_{BC}$, and instrumental limitations prevent the rise from being seen. Indeed, if any rise is seen, it is usually the instrumental response time and not k_{AB}. Occasionally systems do show a measurable rise time, particularly where intermolecular energy transfer is present or when measurements are made in the picosecond time domain.

Equation (4-7) appears complex and is best visualized by considering several limiting cases corresponding to frequently occurring physical systems. For the following arguments, we assume [B]$_0$ = 0. Again this system is analogous to an instantaneous excitation.

We first consider the case where $k_{AB} \gg k_{BC}$. This is a common example for most molecules, in condensed phases, excited to upper excited states. Relaxation to the lowest excited state of the same multiplicity typically occurs with great rapidity ($<10^{-12}$ sec), even though subsequent emission from this state is much slower ($>10^{-9}$ sec). For times much longer than τ_A, Eq. (4-7) reduces to

$$[B] = [A]_0 \exp(-k_{BC}t) \qquad (4-9)$$

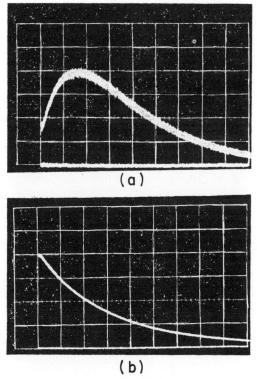

Fig. 4-4. (a) Buildup and decay times of the Eu^{3+} fluorescence intensity in TbPhenAc₃ doped with 10^{-3} molar fraction EuPhenAc₃ at 300°K; phen = 1,10-phenanthroline and Ac = acetate; horizontal scale 500 μusec per main division. (b) Buildup and decay times of the Tb^{3+} fluorescence intensity of TbPhenAc₃ in the same sample as the upper curve; horizontal scale 200 μsec per main division. [From Kleinerman and Choi (1968).]

Thus, if $k_{AB} \gg k_{BC}$ and the luminescence of B is being monitored, the emission looks as if B had been excited directly.

The other limiting case arises if $k_{AB} \ll k_{BC}$. This situation is approximated where a system with a short lifetime is excited by a flash lamp with a long exponential tail. The flash then acts like a long-lived excited state generating the excited emitting species. Then for $t \gg \tau_B$

$$[B] = k_{AB}[A]_0 \exp(-k_{AB}t)/k_{BC} \propto \exp(-k_{AB}t) \tag{4-10}$$

Thus, the luminescence dies out not with the lifetime of the state under study but with the characteristic lifetime of the feeder state which, in this case, is the flash lamp. Therefore, under these conditions, it is impossible to resolve a decay time much shorter than the pump merely by watching the tail of the luminescence decay.

D. PHASE SHIFT MEASUREMENTS OF LIFETIMES

We now describe in greater detail the mathematics of the phase shift method for τ measurements. This method does not use a flash lamp but excites the sample with a sinusoidally varying excitation. The sample lifetime is deduced either from the time or phase shift between the excitation waveform and the sample emission or the degree of modulation of the emission.

Phase shift methods evolved earlier than pulsed methods for short-lifetime measurements because the technology for producing very intense, short light pulses and detecting the weak photoluminescence signals with nanosecond time resolution is relatively recent. The technology for measuring small phase shifts of continuous wave signals even in the presence of high noise levels was well worked out, as was the equipment for sinusoidally modulating the light. Thus, the phase shift method was the workhorse of fluorescence decay time measurements. With the advent of intense pulsed laser sources and single-photon counting methods, phase shift methods have unjustifiably fallen into disrepute. Indeed, of the traditional lifetime measurement methods, the phase shift method is still capable of measuring lifetimes shorter than those measured by any of the conventional methods such as single-photon counting. Further, recent advances in instrumentation and time-resolved spectroscopy suggest continued use of the phase shift approach (Lakowicz, 1983 and references therein).

If a sinusoidally varying excitation source is applied to a luminescent sample, a sinusoidally modulated luminescence is produced. The sample luminescence lags in time the excitation source, or it is phase-shifted. The degree of phase shift is related to τ. This shift arises from the finite sample lifetime which delays luminescence after excitation. The phase shift equations can be derived from Eq. (4-3) as follows: A sinusoidal excitation is given by

$$E(t) = B(1 + M \cos \omega t), \qquad t \geq 0 \qquad \text{(4-11a)}$$

$$E(t) = 0, \qquad t < 0 \qquad \text{(4-11b)}$$

$$\omega = 2\pi f \qquad \text{(4-11c)}$$

where ω is the angular modulation frequency of the excitation, f is the modulation frequency (Hz), B is related to the intensity of the exciting light, and M is the degree of modulation of the exciting light and can assume values of 0–1, which correspond to 0–100% modulation, respectively. Pragmatically, the instantaneous turn-on of $E(t)$ in Eq. (4-11) cannot readily be achieved, but the important modulation characteristics are easily obtained. Substituting Eq. (4-11) into Eq. (4-3) yields $D(t)$ (Appendix B):

$$D(t) = B\{1 - \exp(-t/\tau) + mM[\cos(\omega t - \delta) + m \exp(-t/\tau)]\} \quad \text{(4-12a)}$$

$$m = 1/[1 + (\omega\tau)^2]^{1/2} \quad \text{(4-12b)}$$

$$\delta = \tan^{-1}(\omega\tau) \quad \text{(4-12c)}$$

The expression consists of two groups of terms. The first two terms are the exponential grow-in under conditions of constant intensity and arise from the average intensity in Eq. (4-12) (i.e., the result of $M = 0$). The second group in the square brackets arises from the sinusoidal modulation and gives a sinusoidally modulated part and a decaying exponential.

Figure 4-5 shows $D(t)$ curves for a 10-MHz excitation source ($M = 1$) for lifetimes of 10 and 100 nsec. For the shorter lifetime, the exponential terms quickly die out and a stable sinusoidal $D(t)$ results, which is delayed slightly in time behind $E(t)$. For the longer lifetime, the exponential terms take longer to decay, but eventually $D(t)$ becomes sinusoidal and lags much further behind $E(t)$ than for the short τ. Also, although $D(t)$ is almost fully modulated for $\tau = 10$ nsec, it is only weakly modulated for $\tau = 100$ nsec. This loss of modulation and increased shift is caused by the long residency time in the excited state manifold.

Curves of the form shown in Fig. 4-5 are almost never directly observed. One waits for long times until the transient terms have died out,

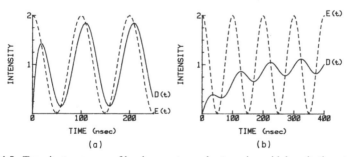

Fig. 4-5. Transient response of luminescent samples to a sinusoidal excitation starting at $t = 0$. In both cases the modulator frequency is 10 MHz. In (a) $\tau = 10$ nsec, and in (b) $\tau = 100$ nsec. The degree of modulation and phase shift at steady state is 84% and 32.14° for (a) and 15% and 80.95° for (b).

which yields

$$D(t) \propto \sin(\omega t - \delta) \qquad (4\text{-}13a)$$

$$m_{obs} = mM \qquad (4\text{-}13b)$$

where δ is the phase shift given in Eq. (4-12) and m_{obs} is the observed degree of modulation of $D(t)$. Here m_{obs} is determined by the intrinsic sample modulation m given by Eq. (4-12b) and the degree of experimental modulation of the source M, which is measurable. To measure lifetimes with a phase shift instrument one measures either the phase shift δ or the degree of modulation m. Then, Eq. (4-12b) or (4-12c) is used to calculate τ. Both methods have been used either separately or simultaneously to complement each other.

Figure 4-6 shows plots of δ and m_{obs} ($M = 1$) versus τ for a 10-MHz modulated source. It is clear that both very long and very short lifetimes relative to the modulation period place severe strains on this lifetime measurement approach. For example, a τ of 0.1 nsec yields $\delta = 0.18°$, and $\tau \sim 0.2$ nsec yields $\delta = 0.36°$; such small phase angles are very difficult to measure with precision or accuracy. Conversely, for very long τ's, δ approaches 90°. A τ of 1000 nsec yields $\delta = 89.1°$, whereas a τ of 5000 nsec yields $\delta = 89.8°$. The discrimination of such small differences is again exceedingly difficult. Another shortcoming is that, for $\tau \gg 1/f$, m becomes small and the measurement becomes increasingly difficult. For example, in the data of Fig. 4-5, $\delta = 32°$ and $m = 84\%$ for $\tau = 10$ nsec, but 80.95° and 15% for $\tau = 100$ nsec.

From Fig. 4-6 it is clear that the optimum frequency for a lifetime measurement using either the δ or m method is when $\tau \sim 1/\omega$. It is, thus, desirable to be able to select several modulation frequencies depending on the range of lifetimes under study. In the past, commonly used piezoelectric-type modulators did not lend themselves to simple frequency changes, as they depended on the resonance frequency of a tuned cavity. At least for molecules absorbing in the red region such as chlorophylls,

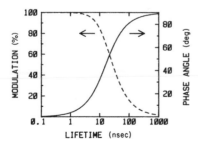

Fig. 4-6. Phase shift (——) and percent modulation (– – –) of a luminescent sample versus sample lifetime. Note the nonlinear lifetime scale. The excitation source is 100% modulated at 10 MHz.

light-emitting diodes have turned out to be excellent sources, and electro-optical modulators can modulate lasers at much shorter wavelengths. Operating frequencies of 100–500 MHz can be achieved yielding low-picosecond time resolution.

The availability of high-intensity light sources and efficient light modulators currently makes the phase shift method an easily implemented, quick, reliable method for obtaining τ's, but it suffers from several disadvantages. First, the necessary equipment can be more extensive and expensive than a primitive flash apparatus. Second, the indirect nature of the measurement can conceal system complexities that flash measurements readily reveal. Specifically, a simple phase shift measurement on a system containing several components with different lifetimes gives no warning of the presence of several species or of a scattered component if only the phase angle at a single frequency is monitored. The flash method, however, yields nonlinear semilogarithmic plots for multiexponential decays. The phase shift method gives no warning, even if the shape of the waveform as well as the phase shift is monitored, because the addition of any number of terms of the form $A_i \sin(\omega t - \delta_i)$ yields a resultant waveform of $B \sin(\omega t - \delta)$. The measured phase shift δ is no longer related to a true lifetime but is a vector average of the contributions of the components with different lifetimes and different amplitudes. By monitoring δ at different frequencies or by monitoring both δ and m, it is possible to detect more complex kinetics. In favorable cases the lifetimes and preexponential factors for a sum of two exponentials can be resolved by the phase shift method. For details see the papers by Birks et al. (1963) and Birks and Munro (1967).

E. STATIC OR ASSOCIATIONAL QUENCHING

The systems described so far all exhibit diffusion-controlled quenching. A variety of systems, however, exhibit an additional quenching pathway. If the donor and the acceptor are in close enough proximity, the donor can be quenched immediately on excitation. Such prompt quenching results only when the donor and quencher are very closely linked with each other, such as in chemically associated pairs, ion pairs, and organized systems such as micelles, bilayers, and monolayers. Studies on static quenching are used for evaluating close associations in organic systems (Birks, 1970), chemical association of metal complexes (Demas and Addington, 1974; Demas et al., 1977a), and ion pairing (Balzani et al., 1975; Demas and Addington, 1976).

The standard way to study such phenomena is by combining steady state intensity and decay time quenching measurements. Static quenching decreases the emission intensity but in the absence of diffusional quenching has no effect on the lifetime of the unassociated excited donors. Thus, it is possible to have a system in which the intensity but not τ decreases as quencher is added. More often, if Q quenches by association, it can also quench by diffusion, and one finds that the lifetime also decreases on adding Q, but not as rapidly as the intensity decreases.

As an example of an associationally quenched system we consider the following example where static and associational quenching are present.

$$D + Q \overset{K_{eq}}{\rightleftharpoons} DQ \tag{4-14a}$$

$$*D \overset{k_1}{\longrightarrow} D + h\nu \quad \text{or} \quad \Delta \tag{4-14b}$$

$$*D + Q \overset{k_2}{\longrightarrow} D + Q + \Delta \tag{4-14c}$$

$$DQ + h\nu \longrightarrow DQ + \Delta \tag{4-14d}$$

Excitation of DQ is assumed to produce no detectable luminescence, feedback of *(DQ) to form *D is absent, and only a one-to-one association complex exists. In Eq. (4-14c) we indicate only catalytic deactivation *D by Q, but all the bimolecular pathways of Eq. (3-1) can be operative, in which case k_2 is given by Eq. (3-1h). For simplicity, we assume $[Q]_0 \gg [D]_0$, where the subscript 0 denotes the formal added concentrations. We note immediately that, in τ measurements, normal Stern–Volmer kinetics are obeyed and

$$(\tau_0/\tau) - 1 = K_{SV}[Q]_0 \tag{4-15}$$

The emission intensity will, however, be affected by two factors: (1) the fraction of free D excited and (2) the amount of *D that is quenched. It can be shown (Birks, 1970; Demas, 1976; Balzani et al., 1975) that the emission intensity ϕ is given by

$$(\phi_0/\phi) - 1 = (K_{SV} + \beta K_{eq})[Q] + \beta K_{SV} K_{eq}[Q]^2 \tag{4-16}$$

where $\beta = 1$ if the solutions are optically dilute and $\varepsilon_{DQ}/\varepsilon_D$ if the solutions are totally absorbing. ε_{DQ} and ε_D are the molar extinction coefficients of DQ and D, respectively, at the excitation wavelength. Equation (4-16) shows that, if only associational quenching is present (i.e., $K_{SV} = 0$), apparently normal intensity Stern–Volmer data result. Thus, on the basis of intensity quenching data alone, one cannot be absolutely sure that quenching is diffusional.

If neither K_{eq} or K_{SV} is zero, nonlinear, upward-curved intensity Stern–Volmer plots result. Because of the symmetry of Eq. (4-16), curve fitting can yield only an ambiguous solution. The use of Eq. (4-15) yields K_{SV} directly, and then βK_{eq} can be extracted from Eq. (4-16). A convenient way to evaluate βK_{eq} is to note that

$$K_{SV}^{app} = [(\phi_0/\phi) - 1]/[Q] \qquad (4\text{-}17a)$$

$$= (K_{SV} + \beta K_{eq}) + \beta K_{SV} K_{eq}[Q] \qquad (4\text{-}17b)$$

where K_{SV}^{app} is the apparent Stern–Volmer constant calculated from intensity data. A plot of K_{SV}^{app} versus [Q] should be a straight line with a slope of $\beta K_{SV} K_{eq}$ and an intercept of $K_{SV} + \beta K_{eq}$. By using K_{SV} from Eq. (4-15), βK_{eq} can be evaluated from the slope and the intercept; the consistency of these two βK_{eq} is a good test of the model.

Figure 4-7 shows a system exhibiting both static and dynamic quenching. In this case $D = Ru(phen)_2(CN)_2$ (phen = 1,10-phenanthroline) with a Ni^{2+} quenching. Ni^{2+} is a relatively poor diffusional quencher of the Ru(II) complex but does associate well by coordination to the CN which acts as a bridging ligand. The result is that most of the observed intensity quenching is by chemical association rather than by diffusion. The best-fit curves were calculated using Eqs. (4-15) and (4-16). Then K_{SV} was obtained directly from τ data [Eq. (4-15)], and βK_{eq} was extracted from the slope and intercept of the K_{SV}^{app}-versus-$[Ni^{2+}]$ plot [Eq. (4-17)].

The important aspect of these two equations is that, using lifetimes and intensity measurements only, it is possible to evaluate equilibrium constants. This approach has been very useful in the evaluation of both

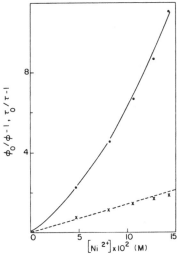

Fig. 4-7. ϕ (●) and τ (x) Stern–Volmer plots for quenching of $Ru(phen)_2(CN)_2$ by Ni^{2+}. The dashed line is the calculated τ plot for $K_{SV} = 14.3 \ M^{-1}$. The solid line is the calculated ϕ fit for $K_{SV} = 14.3 \ M^{-1}$ and $K_{eq} = 12 \ M^{-1}$. [Reprinted with permission from Demas et al. (1977a). Copyright 1977 American Chemical Society.]

chemical association constants and ion pairing constants for a variety of transition metal complex photosensitizers.

We turn now to an alternative way of obtaining both K_{eq} and K_{SV} from a single series of decay time measurements. We assume that we have a delta function excitation and fast detection electronics such that the observed sample response will be a decaying exponential, $K \exp(-t/\tau)$. In the absence of competitive absorption by Q and with a constant-intensity flash, the sample response is given by

$$D(t) = F_D K \exp(-t/\tau) = D_0 \exp(-t/\tau) \qquad (4\text{-}18a)$$

$$(\tau_0/\tau) - 1 = K_{SV}[Q] \qquad (4\text{-}18b)$$

where D_0 is the observed preexponential factor and F_D is the fraction of D present as free D. Equation (4-18b) is the normal decay time form of the Stern–Volmer equation and can be used for direct evaluation of K_{SV}. If associational quenching is absent, F_D will always be unity and D_0 will be independent of [Q]. In the presence of associational quenching, however, F_D will be reduced with increasing [Q], which results in less excitation of free D and a lowering of D_0, permitting calculation of F_D. From F_D versus [Q], K_{eq} is readily calculated.

F. RESONANCE ENERGY TRANSFER

In addition to simple diffusional energy transfer which generally requires contact or near contact of the donor and acceptor, long-range noncontact energy transfer quenching is also possible. This long-range transfer generally involves a dipole–dipole resonance interaction between the donor and acceptor. A necessary condition for resonance energy transfer is an overlap between the donor's emission spectrum and the acceptor's absorption spectrum. With good overlap and strong acceptor absorption, efficient energy transfer can occur at distances of 40 Å or more. Förster (1948, 1949) formulated this current model which is now widely used as a molecular yardstick for measuring distances in polymers and proteins. For extensive discussion of this model the reader is referred to Birks (1970).

We consider here only the problem of uniformly distributed donors and acceptors and the effect of resonance energy transfer quenching on lifetime measurements. Further, only the emission from the donor is considered, although in favorable cases the acceptor emission can also be monitored. Experimentally, this requires that the acceptor be nonlumi-

nescent or that it be experimentally possible to view only the donor luminescence independent of the acceptor emission. Our discussion follows
that of Bennett (1964).

The critical transfer distance R_0 is the separation between the donor,
D, and acceptor, A, at which the rate of energy transfer from *D to A
equals the normal decay rate of *D. Förster theory gives R_0 (cm) from

$$R_0^6 = 9 \times 10^{-25}\beta^2 Q\Omega/n^4 \qquad (4\text{-}19a)$$

$$\Omega = \int f_D(\bar{\nu})\varepsilon_A(\bar{\nu}) \, d\bar{\nu}/\bar{\nu}^4 \qquad (4\text{-}19b)$$

where β^2 is an orientation factor which for randomly oriented nonrotating
donors and acceptors is 0.475 (Birks, 1970), n is the solvent refractive
index, Q is the absolute photon yield of unquenched donor, Ω is the
overlap integral for the donor emission with the acceptor absorption, $f_D(\bar{\nu})$
is the corrected luminescence spectrum of the donor versus the wave
number $\bar{\nu}$ and is normalized to unit area, and $\varepsilon(\bar{\nu})$ is the molar extinction
coefficient of the acceptor versus $\bar{\nu}$. The overlap integral is evaluated over
the entire donor emission. The critical concentration C_A^0 of the acceptor is
given by

$$C_A^0 = 4 \times 10^{-10} n^2/\beta(Q\Omega)^{1/2} \qquad (4\text{-}20)$$

The decay curve for an impulse excitation source and a translationally
frozen system is given by

$$D(t) = D(0) \exp(-t/\tau) \exp[-2\gamma(t/\tau)^{1/2}] \qquad (4\text{-}21)$$

where τ is the unquenched lifetime and γ is given by

$$\gamma = C_A/C_A^0 \qquad (4\text{-}22)$$

where C_A is the actual quencher concentration. Bennett (1964) has reported studies of the dynamics of donor decay in systems involving Förster energy transfer between bis(hydroxyethyl)-2,6-naphthalenedicar-
boxylate to the dye Sevron Yellow G2 and from pyrene to Sevron Yellow
L. Figure 4-8 shows the observed and predicted decay curves for the
pyrene system as a function of the acceptor concentration. In this case the
necessary R_0 required to fit the experimental decay curves are ~42–46 Å
which agree well with the 39 Å predicted by Eq. (4-19).

The shapes of the curves of Fig. 4-8 are quite interesting. Initially, for
quenched samples, the decay is steep, but at long enough times the slope
of the $\log[D(t)]$-versus-t plots linearize with lifetimes equal to the unquenched lifetime. The higher the quencher concentration, the steeper the

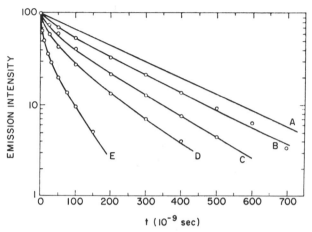

Fig. 4-8. Decay curves for pyrene under conditions of Förster energy transfer quenching by Sevron Yellow L in a solid polymer film. Calculated and observed decay curves with the acceptor concentrations in units of 10^{-3} moles/liter): A, 0; B, 0.63; C, 2.4; D, 4.3; E, 9.4. [From Bennett (1964).]

initial slope and the longer the period before the slope equals that of the unquenched sample. This is quite different from normal Stern–Volmer quenching where the lifetime shortens with increasing quencher concentration but the decays remain exponential. This odd time dependence of Förster quenching arises because quenchers very near the donor quench more rapidly than those far from the donor. Thus, one observes a rapid fall off of excited state concentration due to the rapid deactivation of donors situated near quenchers. At long enough times, the only donors left are those that are too far from a quencher to be deactivated by the Förster mechanism and, thus, decay with their unquenched lifetimes. This sort of behavior requires, however, that the relative positions of donors and acceptors do not change during the excited state lifetime.

It should also be pointed out that Förster energy transfer does not require a fluorescence. Triplet states exhibiting phosphorescence can likewise exhibit Förster energy transfer to suitable acceptors with the same temporal dependence. Because the donor emission is spin-forbidden and slow, however, the energy transfer process is also correspondingly slow. For example, Bennett et al. (1964) reported triplet-to-singlet energy transfer from phenanthrene-d_{10} to rhodamine B. The shapes of the decay curves versus the rhodamine B concentration were virtually identical to those of Fig. 4-8, except that the time scale was 0–50 sec rather than in the submicrosecond range.

G. COUPLED SYSTEMS

1. General Solution

A very important class of systems involves coupled reactions:

$$A \underset{k_{BA}}{\overset{k_{AB}}{\rightleftharpoons}} B \tag{4-23}$$

$$k_A \downarrow \qquad \downarrow k_B$$

products products

This is similar to a serial decay scheme, except that B can now feed back to A. Systems that fall into this category include two nearly isoenergetic excited states, excimers and exciplexes, and excited state acid–base reactions. We discuss the general solution initially and excimer and excited state acid–base systems later.

The differential equations for Eqs. (4-23) are

$$d[A]/dt = -(k_A + k_{AB})[A] + k_{BA}[B] \tag{4-24a}$$

$$d[B]/dt = +k_{AB}[A] - (k_B + k_{BA})[B] \tag{4-24b}$$

We assume that the boundary value conditions are that, at $t = 0$, $[A] = [A]_0$ and $[B]_0 = 0$. These coupled first-order homogeneous differential equations have the following solutions (Appendix C):

$$[A] = [A]_0[g_1 \exp(-\gamma_1 t) - g_2 \exp(-\gamma_2 t)] \tag{4-25a}$$

$$g_1 = (X - \gamma_2)/(\gamma_1 - \gamma_2)$$

$$g_2 = (X - \gamma_1)/(\gamma_1 - \gamma_2)$$

$$[B] = [A]_0[k_{AB}/(\gamma_1 - \gamma_2)][\exp(-\gamma_2 t) - \exp(-\gamma_1 t)] \tag{4-25b}$$

where

$$\gamma_1 = \{(X + Y) + [(X - Y)^2 + 4k_{AB}k_{BA}]^{1/2}\}/2 \tag{4-26a}$$

$$\gamma_2 = \{(X + Y) - [(X - Y)^2 + 4k_{AB}k_{BA}]^{1/2}\}/2 \tag{4-26b}$$

$$1/\tau_1 = \gamma_1 \tag{4-26c}$$

$$1/\tau_2 = \gamma_2 \tag{4-26d}$$

$$X = k_A + k_{AB} \tag{4-26e}$$

$$Y = k_B + k_{BA} \tag{4-26f}$$

In this general case A or B does not decay exponentially but exhibits complex decay kinetics. In both cases, however, the decay curves are the sum of two exponential terms with different preexponential factors; but the characteristic rate parameters γ_1 and γ_2 for the exponential terms are the same for the time dependence of A and of B. Further, if we define lifetimes τ_1 and τ_2 in Eqs. (4-26c) and (4-26d), the concentrations of Eqs. (4-25) are the sum of two exponential decays with associated apparent lifetimes τ_1 and τ_2. In general, τ_1 and τ_2 do not represent the lifetimes of individual components. We present below examples of the complexity of these kinetics.

The series decay of Section C is actually a limiting case of the scheme of Eq. (4-25) with $k_{BA} = 0$. This is readily shown by substitution which yields $\tau_1 = (k_A + k_{AB})^{-1}$ and $\tau_2 = k_B^{-1}$.

Figure 4-9 demonstrates the effect of adding B-to-A coupling to a simple series decay. For simplicity, we choose $k_A = 0$ and $k_{AB} = 2k_B = 5 \times 10^7 \text{ sec}^{-1}$. Plots of [A] and [B] versus t are shown for values of k_{BA}/k_{AB} ranging from $\frac{1}{8}$ to 4 along with the uncoupled case of $k_{BA} = 0$. Several important results follow. Even for relatively small degrees of feedback ($\frac{1}{8}$), there is substantial distortion of the [A] and [B] curves relative to a simple serial decay with no feedback. For increasing amounts of feedback, the [A] curve becomes increasingly nonexponential and is clearly visually nonexponential for a k_{BA}/k_{AB} of 1. For greater feedback, the [A] and [B] curves versus time become nearly linear at longer times.

Another important case arises of $k_B = 0$. Assuming $k_A = 5 \times 10^7 \text{ sec}^{-1}$ and $k_{AB} = 20k_A$, we plot a family of decay curves in Fig. 4-10; the logarithmic plots are included to permit easier viewing of the low concentrations. Again, the decays of A and B are complexly related to k_{BA}/k_{AB}.

Figure 4-10 reveals an extremely interesting and important result. If k_{AB} and k_{BA} are much greater than k_A, then the exchange between A and B

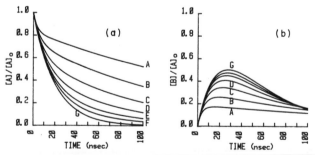

Fig. 4-9. Decay curves for A and B in an A $\rightleftarrows$ B $\rightleftarrows$ C system involving coupling of B to A. In all cases $k_{AB} = 5 \times 10^7 \text{ sec}^{-1}$, $k_B = 2.5 \times 10^7 \text{ sec}^{-1}$, and $k_A = 0 \text{ sec}^{-1}$. Plots in (a) are for [A]/[A]$_0$ and in (b) for [B]/[A]$_0$. The letters correspond to the following different values of k_{BA}/k_{AB}: A, 4/1; B, 2/1; C, 0.99/1; D, 1/2; E, 1/4; F, 1/8; G, 0/1. Case G is the curve for no feedback from B to A.

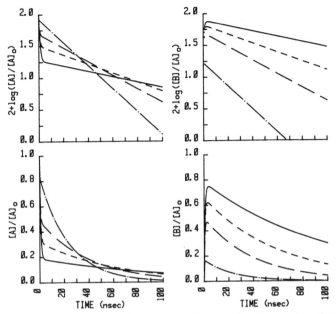

Fig. 4-10. Decay curves for A and B in a coupled system. In all cases, $k_A = 5 \times 10^7 \text{ sec}^{-1}$, $k_{AB} = 10^9 \text{ sec}^{-1}$, and $k_B = 0$. The curves correspond to the following values of k_{AB}/k_{BA}: ——, 4/1; ---, 2/1; ————, 1/1; —·—·—, 1/5.

is much faster than the relaxation of A and a pseudoequilibrium exists in the A/B ratio. After a short transient settling period the decays of both A and B are exponential with the same lifetimes. Birks (1970) refers to this as "dynamic equilibria." This common lifetime is long, however, compared to $1/k_A$ which would be the lifetime of A in the absence of feedback from B. This situation arises because at long times [A] and [B] are in a pseudoequilibrium. The only path open for a return to the ground state of A or B is by k_A, and the rate is $k_A[A]$. As k_{AB} increases relative to k_{BA}, after the induction period, there is less A present at any instant and the rate of loss of A decreases. [A] and [B] track each other, because as [A] decreases, the exchange between A and B is so rapid that the pseudoequilibrium between A and B is maintained.

Systems corresponding to this limiting case include type-E thermally populated delayed fluorescence (Parker, 1968). The excitation is trapped in a nonluminescent triplet state which is thermally coupled to a higher-level fluorescent singlet state. The decay of the fluorescence then tracks the decay of the long-lived triplet state which functions as an energy storage state. In actual type-E fluorescence, k_B is not zero, and only a relatively small portion of the excited molecules actually decay through A.

2. Excimer and Exciplex Formation

A class of extremely important coupled reactions involve excimers and exciplexes. Many molecules will not associate or react with each other in the ground state. Once one of them is in an excited state, however, it will undergo reactions with other normally nonreacting ground state species to form new transient excited state species. This is not a permanent chemical reaction, and when the new excited species decay, the original ground state species are formed. If the reactant is a ground state molecule of the same type, the new species is an excimer (*excited state dimer*). If the two species are different, they are called exciplexes (*excited state complexes*), although if the two species are similar, the term "mixed excimer" is sometimes used.

The first and most well-known example of an excimer is pyrene. Pyrene has absolutely no ground state association with itself, but in the excited state efficiently forms excimers. Figure 4-11 shows the concentration dependence of the pyrene luminescence. At low concentrations only the high-energy structured monomer emission exists. At higher concentrations excited state reaction to the excimer occurs to give the lower-energy excimer emission. The lower-energy emission is readily shown to arise by a slow step by means of transient lifetime measurements. The excimer grows in after a delay following an instantaneous excitation.

The equations describing excimer or exciplex systems are given by

$$\text{*M} + \text{N} \underset{k_{\text{DM}}}{\overset{k_{\text{MD}}}{\rightleftharpoons}} \text{*(MN)} \tag{4-27}$$

$$k_{\text{M}} \downarrow \qquad\qquad k_{\text{D}} \downarrow$$

$$\text{M} + h\nu_1 + \text{N or } \Delta \qquad \text{M} + \text{N} + h\nu_2 \text{ or } \Delta$$

where *M is the excited species, N is the ground state reactant, and *(MN) is the excimer or exciplex. If $M = N$, then we have excimer formation, whereas if the two are different, we have mixed excimer or exciplex formation.

The scheme of Eq. (4-27) yields the following differential equations:

$$d[\text{*M}]/dt = -(k_{\text{M}} + k_{\text{MD}}[\text{N}])[\text{*M}] + k_{\text{DM}}[\text{*MN}] \tag{4-28a}$$

$$d[\text{*MN}]/dt = +k_{\text{MD}}[\text{N}] - (k_{\text{D}} + k_{\text{DM}})[\text{*MN}] \tag{4-28b}$$

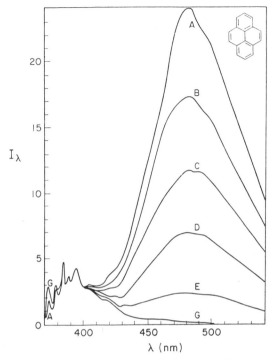

Fig. 4-11. Fluorescence spectra of pyrene solutions in cyclohexane. Intensities have been normalized to the high-energy structured monomer emission. A, 10^{-2} M; B, 7.75×10^{-3} M; C, 5.5×10^{-3} M; D, 3.25×10^{-3} M; E, 10^{-3} M; G, 10^{-4} M. [Reprinted with permission from Birks and Christophorou (1963). Copyright 1963, Pergamon Press, Ltd.]

Comparison of Eqs. (4-28) and (4-24) shows them to be identical if we make the following substitution

$$[A] = [*M], \qquad [B] = [*MN]$$

$$k_A = k_M, \qquad k_B = k_D \qquad (4\text{-}29)$$

$$k_{AB} = k_{MD}[N], \qquad k_{BA} = k_{DM}$$

Thus, the solutions for Eqs. (4-28) are Eqs. (4-25) using the relations of Eq. (4-29). This solution is valid as long as one is not working with such high-excitation flux conditions that the concentration of M or N is seriously depleted. In other words, one must be working under pseudo-first-order conditions for the reactions of *M.

The amount of work that has been done on excimers and exciplexes is nothing short of phenomenal. For a summary of many results and a

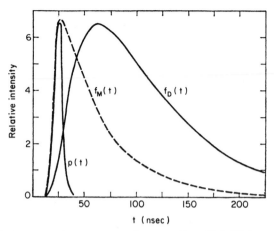

Fig. 4-12. Fluorescence decay curves of pyrene (5×10^{-3} M) in cyclohexane. The monomer decay $f_M(t)$ and the excimer $f_D(t)$ are resolved from each other by monitoring at their respective emission maxima (Fig. 4-11). The excitation pulse $p(t)$ is not quite short enough to be treated as a delta function for the early portions of $f_M(t)$. [From Birks *et al.* (1963).]

method of treating both steady state and transient data, see the book by Birks (1970).

Figure 4-12 shows a typical excimer transient. In this case, both M and N are pyrene. Pyrene has been a particularly well-studied system because its excited state lifetime is long (>100 nsec), which both facilitates excimer formation and simplifies transient measurements. Note that, in this case, the excitation is not short enough to represent a zero duration pulse and distorts the leading edge of both the monomer and the excimer decays. Modern single-photon data are in fact, much cleaner because of the shorter temporal resolution. Deconvolution (Chapter 8) methods permit the removal of flash artifacts.

3. Excited State Acid–Base Reactions

Much as the excited state properties of a monomeric species can be greatly different from the ground state properties as in excimer or exciplex formation, acid–base properties can also change greatly in the excited state. It is not uncommon for an acid, HD, to be four to five orders of magnitude more or less acidic in the excited state than in the ground state, and seven to nine orders of magnitude is not uncommon (Parker, 1968). For comparison, the pK_a difference between the weak acetic acid and the second dissociation of the moderate-strength sulfuric acid is only

3 pK units. These large differences between the ground and excited state pK values for the same molecule arise because the excited species can have an electron distribution entirely different from that of the ground state and, thus, can be thought of as an entirely different chemical species.

Excited state acid–base reactions can be described by the following set of reactions:

$$*Z^- + H^+ \underset{k_{HZZ}}{\overset{k_{ZHZ}}{\rightleftharpoons}} *(HZ) \qquad (4\text{-}30)$$

$$k_Z \downarrow \qquad\qquad\qquad k_{HZ} \downarrow$$

ground state products   ground state products

Unlike the excimer case it is now possible, depending on [H$^+$], to have the ground state of the system consist of essentially all Z$^-$, all HZ, or a mixture of both Z$^-$ and HZ. Depending on pH the initial composition of the solution on pulsed excitation consists of all *Z$^-$, all *(HZ), or a mixture of *Z and *(HZ). The two limiting cases of all *Z$^-$ or all *(HZ) have as their solution Eq. (4-25) after proper redefinition of terms.

If [*(HZ)]$_0$ = 0, then Eq. (4-25) uses the following definitions

$$[*Z^-] = [A], \qquad [*(HZ)] = [B]$$

$$k_A = k_Z, \qquad k_B = k_{HZ} \qquad (4\text{-}31)$$

$$k_{AB} = k_{ZHZ}[H^+], \qquad k_{BA} = k_{HZZ}$$

If [*Z$^-$]$_0$ = 0, then the necessary definitions are

$$[*(HZ)] = [A], \qquad [*Z^-] = [B]$$

$$k_A = k_{HZ}, \qquad k_B = k_Z \qquad (4\text{-}32)$$

$$k_{AB} = k_{HZZ}, \qquad k_{BA} = k_{ZHZ}[H^+]$$

For the general case where neither [*HZ]$_0$ nor [*Z$^-$]$_0$ is zero, recourse must be made to the general solution. This can most easily be obtained by noting that the solution for the general case is just the sum of the solutions for the two limiting cases with the appropriate experimental [*HZ]$_0$ and [*Z$^-$]$_0$.

It should be pointed out that excited state proton transfer reactions are extremely common. They occur with a variety of common organic molecules (e.g., naphthol), biological probes and biological systems, laser dyes, and metal complexes (Peterson and Demas, 1976, 1979). This is particularly true when one recognizes that a change of five orders of magnitude in acidity can convert a moderate-strength acid into a super-acid or a strong acid into a weak one. The reader is referred to the literature for further information on this topic. In particular, we wish to stress that proper treatment of such data can be extremely complex. Unlike excimer and exciplex systems where there are frequently emission regions where only *M and only *(MN) emit, it is common to have extensive overlap in systems exhibiting excited state acid–base reactions. An excellent example of the complexities of such data treatment is given by the exhaustive study on 2-naphthol by Laws and Brand (1979).

H. STEADY STATE APPROXIMATION AND EXCITED STATE EQUILIBRIA

The steady state approximation (SSA) is a commonly used method of simplifying complex kinetic behavior. Application of the SSA can often result in considerable simplification of the kinetics. However, arbitrary use of the SSA can yield totally incorrect equations and interpretations. We conclude this chapter with a discussion of the assumption of excited state equilibria. We then more critically analyze this assumption by the SSA and demonstrate how uncritical use of the SSA can lead to severe errors. For purposes of the current analysis, we consider only the relatively simple system of Eqs. (4-23) because we have a rigorous closed form solution for comparison with the approximate solutions. Our conclusions are readily generalizable to more complex systems.

In the coupled systems of Eq. (4-23) a frequently used simplification is to assume that an excited state equilibrium exists between A and B. That is, one assumes that the reactants A and B can be treated as though they were in equilibrium in the reaction

$$A \underset{k_{BA}}{\overset{k_{AB}}{\rightleftharpoons}} B \qquad (4-33)$$

One can then define an excited state equilibrium constant $*K_{eq}$ in the traditional kinetic sense by

$$[B]/[A] = *K_{eq} = k_{AB}/k_{BA} \qquad (4-34)$$

This frequently yields substantial simplifications, but it relies on an important condition: The excited state equilibrium must be established much faster than A or B decay by the k_A or k_B route. In the absence of the k_A and k_B pathways, equilibrium is approached exponentially with a rate constant $k_{AB} + k_{BA}$ (Laidler, 1965). Therefore, if k_{AB} and k_{BA} are much greater than k_A and k_B, a pseudoequilibrium for [A] and [B] can be established, and one can use Eq. (4-34) for calculating [B]/[A].

An important case where an excited state equilibrium is suitable arises in thermal equilibration of the excited state manifold. For example, in condensed media, equilibrium within the excited-state singlet manifolds is typically complete within a few picoseconds following excitation to upper singlet states. For an emission occurring on a nanosecond time scale, equilibration of the emitting manifold can be considered essentially perfect.

Another important class of systems that can satisfy Eq. (4-34) is excited state acid–base reactions. If the forward and backward rates are fast enough, a pseudoequilibrium will exist. A normal titration curve of emission of the protonated or the unprotonated form versus pH then results. The inflection point of such a curve then occurs at pK^*. β-Naphthylamine is a system that generally exhibits excited state equilibrium (Förster, 1960; Parker, 1968).

If, however, one works in a pH range where [H$^+$] is too low to react rapidly with *Z$^-$ before it can decay, incomplete equilibration results. The resultant titration curves with such a system might look relatively normal, but the apparent pK^* is caused by kinetic limitations rather than thermodynamic ones. β-Naphthol is an example of a system that has incomplete equilibration (Parker, 1968).

We now consider what happens if k_A and/or k_B becomes significant compared to k_{AB} and k_{BA}. Figure 4-13 shows typical [B]/[A] curves versus time. In all curves $k_{AB} = k_{BA} = 1$ sec^{-1} and [B] $= 0$ at $t = 0$. For simplicity, either k_A or k_B is zero. The value for the nonzero k_A or k_B is indicated to the right of each curve. The curve labeled 0.00 corresponds to $k_A = k_B = 0$ and yields the expected equilibrium ratio of 1 [Eq. (4-34)]. The curves above this one have $k_B = 0$, and the curves below it have $k_A = 0$. The approach to a pseudo–excited state equilibrium is clearly evident in all cases. However, even if k_A or k_B is only 25% of k_{AB} and k_{BA}, severe errors in the pseudo–equilibrium value calculated from Eq. (4-34) result. Even when there are errors, the equilibration time is comparable to that in the true equilibrium case.

That [B]/[A] quickly achieves a steady state might suggest to the incautious that the widely used steady state approximation of kinetics may

Fig. 4-13. Approach to pseudoequilibrium for a coupled system of Eq. (4-23) for different rate constants. In all cases $k_{AB} = k_{BA} = 1$ sec^{-1}. The curve labeled 0.00 corresponds to $k_A = k_B = 0$ or simple chemical equilibration. For curves below this, $k_A = 0$ and the indicated value is k_B. For curves above it, $k_B = 0$ and the indicated value is k_A.

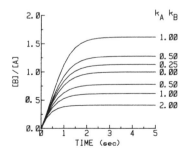

be used. The steady state approximation assumes that, after an initial transient period, the rate of change in the concentrations of reactive intermediates present at low concentrations can be set equal to zero. Actually, what is being assumed is that the rate of change in this intermediate is much smaller than for other components. Thus, this approximation might be useful for calculating the pseudo-steady state [B]/[A]'s of Fig. 4-13. We shall now demonstrate with a slightly different example that this steady state approximation is totally inappropriate under the current conditions.

Consider the specific case where $k_A = 0$. If we assume that we can use the steady state approximation for B, we have

$$d[B]/dt = k_{AB}[A] - (k_{BA} + k_B)[B] = 0 \qquad (4\text{-}35)$$

which yields at steady state

$$[B]/[A] = k_{AB}/(k_{BA} + k_B) \qquad (4\text{-}36)$$

Note that, if $k_B = 0$, the equilibrium case of Eq. (4-34) results. If we have $k_{BA} = 1$ sec^{-1} and $k_B = 3k_{BA}$, for $k_{AB} = 1$ sec^{-1} the steady state approximation yields [B]/[A] = 0.25, whereas the rigorously correct solution [Eq. (4-25)] yields 0.434 or a 42% error. The reason for this failure of the steady state assumption is clear if one looks at [A] and [B]. Both concentrations are comparable at steady state. Thus, neither $d[A]/dt$ nor $d[B]/dt$ can be much smaller than the other. This precludes setting either equal to 0 as was done to derive Eq. (4-36). When, then, is the steady state approximation correct? The answer is when one of the species is present in very low concentration relative to the others.

In the above example with $k_B = 3k_{BA}$ we can force the steady state concentration of [B] to lower and lower values by holding k_B and k_{BA} fixed while decreasing k_{AB}. If k_{AB} becomes small enough, [B] will approximate a low-concentration reactive intermediate and the steady state assumption will hold. To demonstrate this effect graphically, we hold $k_{BA} = 1$ sec^{-1}, $k_B = 3k_{BA}$, and vary k_{AB}. Instead of plotting [B]/[A], we normalize

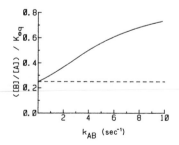

Fig. 4-14. Comparison of the steady state (---) and actual (——) pseudoequilibrium position for a coupled system with $k_{BA} = 1$ sec^{-1} and $k_B = 3k_{BA}$ versus k_{AB}.

it $[([B]/[A])/K_{eq}]$. For an excited state system K_{eq} is $^*K_{eq}$. Using the steady state approximation

$$([B]/[A])/K_{eq} = 1/[1 + (k_B/k_{BA})] \qquad (4\text{-}37)$$

This normalized ratio equals 0.25 for the current example.

Figure 4-14 is a plot of the true dependence of the normalized ratio (solid line) and the steady state approximation (dashed line). Clearly, for relatively large k_{AB}, the steady state approximation is a disaster differing from the correct value by a factor of 3. Only for very small k_{AB} does the true curve approach the steady state prediction. For the steady state approximation to be accurate to 1%, k_{AB} must be 0.05 k_{BA}. The warning is clear: Use caution in applying the steady state approximation and verify after completing the calculation that the component to which it applied was a suitable one.

Least Squares Data Reduction

A. STATEMENT OF LEAST SQUARES METHOD

Most of the previously described systems are characterized by relatively simple equations. Generally, given specific parameters, it is simple to calculate an expected curve. Frequently, it is much more difficult to extract the parameter values from experimental data. A number of the systems can be handled easily by graphical methods with visual fitting; the linear fit to the semilogarithmic plot versus time of an exponential decay and simple Stern–Volmer quenching are examples. Visual fitting is not the most accurate method for complex defining equations and may not even be possible (e.g., excimer or Förster kinetics).

The method of least squares supplies a relatively simple, general approach to the fitting of curves where the defining equations for the data set are known and the values of the parameters giving the best fit are desired. In this chapter we first develop the method of linear least squares and discuss some of its pitfalls. The linearization of functions and the weighting factor problem are discussed. The least squares approach is then extended to the fitting of nonlinear and nonlinearizable equations such as the sum of exponentials. Several different nonlinear fitting methods and the estimation of statistical error in the final parameter estimates will be covered.

We assume the data are ordered pairs (x_i, y_i) and that the data are fit by $y = F(x)$. The least squares method seeks the parameter values in $F(x)$

that minimize chi-square χ^2:

$$\chi^2 = \sum_{i=1}^{N} w_i R_i^2 \tag{5-1a}$$

$$R_i = y_i - F(x_i) \tag{5-1b}$$

$$w_i = (1/\sigma_i^2) \tag{5-1c}$$

where w_i is the weighting factor, σ_i is the standard deviation of the observed y_i, and $F(x_i)$ is the modeling function evaluated at x_i. The residuals R_i are the differences between the observed data and the modeling function. The summation is over the N data points to be fit. The weighting factor emphasizes or weights more heavily the more accurately known points. The proper weighting factors are given by Eq. (5-1c) (Bevington). Equation (5-1) assumes that all errors are concentrated in the y_i, and this is frequently valid.

Statistical information is not always available, and it is then common to assume that all w_i are equal. For simplicity, the w_i are all set equal to unity. This use of $w_i = 1$, rather than some other constant, has no effect on the values of the parameters that minimize χ^2; the use of different constants merely scales the sum. We derive expressions using weighting factors; the unweighted case is obtained with all $w_i = 1$.

We first consider fitting data to a straight line:

$$F(x) = y = a_1 + a_2 x \tag{5-2}$$

The least squares method selects a_1 and a_2, which minimize χ^2:

$$\chi^2 = \Sigma w_i R_i^2 = \Sigma w_i [y_i - (a_1 + a_2 x_i)]^2 \tag{5-3}$$

For simplicity, we have omitted the indexes on the summations that run from 1 to N. Consider specifically the Stern–Volmer lifetime quenching data set of Fig. 3-3 where the y_i are defined as $(\tau_0/\tau) - 1$ and the x_i are [$HgCl_2$]. Figure 5-1 is a contour map of the χ^2 surface plotted versus a_1 and a_2; all $w_i = 1$. The a_1 and a_2 giving a minimum χ^2 could be estimated graphically, but these calculations are exceedingly tedious. From Fig. 5-1, we estimate $a_1 = 112 \; M^{-1}$ and $a_2 = 1.10$. A much less complex process arises if one considers the necessary conditions for a minimum in the χ^2 surface; the partial derivatives of χ^2 with respect to both a_1 and a_2 must be simultaneously zero:

$$(\partial\chi^2/\partial a_1)_{a_2} = 0 \tag{5-4a}$$

$$(\partial\chi^2/\partial a_2)_{a_1} = 0 \tag{5-4b}$$

Actually Eqs. (5-4) guarantee only that one is at a minimum, a maximum, or a saddle point in the surface; but a linear fit yields only a minimum. In

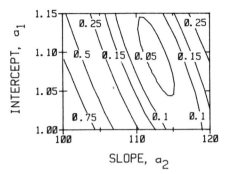

Fig. 5-1. Contour map of χ^2 surface for the decay time Stern–Volmer quenching data of Fig. 3-3 ($[(\tau_0/\tau) - 1]$ versus $[HgCl_2]$). Here a_1 and a_2 correspond to the intercept and slope, respectively. Each curve corresponds to a contour of constant χ^2 with the indicated value.

general, however, the reduction method must eliminate such ambiguities. Further, complex nonlinear fitting can give multiple minima, and the fitting method must discern between the desired global rather than local minimum.

Evaluating the necessary partial derivatives in Eq. (5.4) and rearranging yields the two equations

$$(\Sigma w_i)a_1 + (\Sigma w_i x_i)a_2 = \Sigma w_i y_i \tag{5-5a}$$

$$(\Sigma w_i x_i)a_1 + (\Sigma w_i x_i^2)a_2 = \Sigma w_i x_i y_i \tag{5-5b}$$

Equations (5-5) are two linear equations in the two unknowns a_1 and a_2. These equations are the system's "normal" equations and are readily solved by elimination, determinants, or matrix methods to yield

$$a_1 = (\Sigma w_i x_i^2 \Sigma w_i y_i - \Sigma w_i x_i \Sigma w_i x_i y_i)/\Delta \tag{5-6a}$$

$$a_2 = (\Sigma w_i \Sigma w_i x_i y_i - \Sigma w_i x_i \Sigma w_i y_i)/\Delta \tag{5-6b}$$

$$\Delta = \Sigma w_i \Sigma x_i^2 - (\Sigma w_i x_i)^2 \tag{5-6c}$$

These values of a_1 and a_2 yield a minimum χ^2. Using Eq. (5-6) for the data of Fig. 5-1 with all $w_i = 1$ yields $a_1 = 112.7 \ M^{-1}$ and $a_2 = 1.094$, which agree well with the graphical estimates.

We have made the important assumption that the weighting factors were constant, which is equivalent to assuming that the errors in the y_i are constant (i.e., σ_i = constant). This last assumption can lead to significant errors (vide infra).

B. LINEARIZABLE COMPLEX FUNCTIONS

In the previous example, we linearized the nonlinear data set of τ versus [HgCl$_2$] by using $y = (\tau_0/\tau) - 1$ and $x = $ [HgCl$_2$]. Many other nonlinear functions are also readily linearizable by a similar substitution of variables.

The static-dynamic quenching problem of Section 4.E is linearized by plotting the calculated apparent intensity Stern–Volmer quenching constant K_{SV}^{app} versus [Q] and using $y = K_{SV}^{app}$ and $x = $ [Q] in Eqs. (5-6). The intercept then equals $K_{SV} + \beta K_{eq}$, and the slope equals $K_{SV}\beta K_{eq}$. The exponential decay of Eq. (3-4c) is linearized [Eq. (3-4a)] using $y = \ln[D(t)]$ and $x = t$. From Eqs. (5-6), $a_1 = \ln[D_0]$ and $a_2 = (-1/\tau)$. This is, of course, the standard method of evaluating exponential lifetimes.

C. LEAST SQUARES WITH WEIGHTING FACTORS

The common and apparently successful use of unweighted least squares fitting to many functions can produce a false sense of security. Consider, for example, the data of Figure 5-2 given by $D(t) = (10 \text{ V})$ $\exp(-t/40 \text{ nsec})$ with added constant Gaussian noise. The noise has a constant 0.100-V standard deviation. This decay might result from a high-quality signal and an amplifier with amplitude-independent noise. These data appear reasonable for data reduction. An unweighted linear least squares fit of the $\ln[D(t_i)]$-versus-t_i data over the 0–100-nsec range yields $D_0 = 9.946$ and $\tau = 40.18$ nsec. In view of the low noise level and the 100

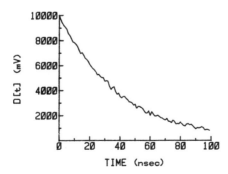

Fig. 5-2. Simulated exponential decay curve with a constant noise level. The noise-free curve is given by (10 V) $\exp(-t/40)$, and the added noise is Gaussian with a standard deviation of 0.1 V.

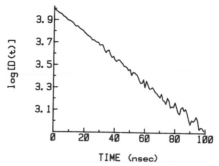

Fig. 5-3. Semilogarithmic plot of the data of Fig. 5-2.

points used in the fit, these results correspond rather poorly to the noise-free $D_0 = 10$ V and $\tau = 40.00$ nsec. The reason for this poor fit is revealed in the semilogarithmic data plot (Fig. 5-3); although quite noise-free at short times, $\ln[D(t)]$ is quite noisy at longer times and, especially for $t \geq$ ~70, exceeds what one might have expected. The poor fit arises because the poor data at long times are weighted the same as the superior data at early times.

The solution is to weight less heavily the less reliable data in the minimization of Eq. (5-6). Proper determination of the w_i is not as obvious as it might seem.

For our example, even though the σ_i on $D(t)$ are constant, the w_i in Eq. (5-6) are not constant; see Fig. 5-3. This difference arises because y is not $D(t)$, but $\ln[D(t)]$, and the noise levels on $\ln[D(t)]$ and $D(t)$ differ. To quantitate this we ask what effect a small change or uncertainty in $D(t)$ has on y. For $y = \ln[D(t)]$,

$$dy = d \ln[D(t)] = [1/D(t)] \, dD(t) \tag{5-7}$$

where dy and $dD(t)$ are differential changes in y and $D(t)$, respectively. Small fractional changes in $D(t)$ [i.e., $dD(t)/D(t)$] produce $D(t)$-dependent changes in y (i.e., dy). For example, assuming we can use incremental for differential changes, if $D(t) = 10$ and $dD(t) = 10^{-1}$, dy is 10^{-2}, but if $D(t) = 1.0$, then dy is 0.1 or 10 times larger. The signal-to-noise ratio (S/N) for $\ln[D(t)]$ improves with increasing $D(t)$. Figures 5-2 and 5-3 clearly show this effect. The uniform $D(t)$ noise level yields barely visible noise in Fig. 5-3 at short times but makes the data almost worthless for fitting at longer times. Thus, in an unweighted fit the low S/N low-intensity points dispro-portionately affect the fit.

If we approximate the standard deviation in $D(t)$ and in y by $dD(t)$ and dy, we obtain

$$\sigma_y = [1/D(t)] \, \sigma_D \tag{5-8}$$

The correct w_i for the semilogarithmic plot are

$$w_i = 1/\sigma_i^2 = D(t_i)^2/0.01 \tag{5-9}$$

For the example, the slope and intercept yield $D_0 = 10.00$ and $C = 40.07$ nsec, which agree much more closely with the correct answers. If these calculations were repeated many times, the weighted fits would typically be more accurate and more precise than the unweighted fits.

Clearly, indiscriminate unweighted least squares fitting of transformed linearized functions must be done with caution. At the very least representative linearized plots and fits should be inspected. A properly weighted fit is always the best approach.

In the general case where the linearized function is $y = F(x)$, we have

$$\sigma_y = [dF(x)/dx]\,\sigma_x \tag{5-10}$$

We consider another example of the adverse effects of improper weighting. Figure 5-4 shows a simulated decay from an SPC experiment. In this case σ_i is given by

$$\sigma_i = D(t_i)^{1/2} \tag{5-11}$$

where $D(t_i)$ are the number of counts in each data channel. The semilogarithmic plot (Fig. 5-4b) shows that the data at long times have a poorer S/N and should be weighted less. The effect is smaller than before, and a wider range of $D(t)$ is necessary to show it.

To test the effects of weighting and not weighting, we calculated 100 synthetic decay curves of the form $K \exp(-t/\tau)$ with added SPC noise. The weighted and unweighted least squares fits were made using the $\ln[D(t)]$-versus-t plots. The average derived preexponential factor K and lifetime τ, as well as the standard deviation for each set of decay curves,

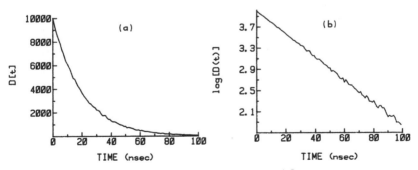

Fig. 5-4. Decay curve (a) and semilogarithmic plot (b) of a simulated single-exponential decay curve $10^4 \exp(-t/20)$. Noise conforming to a single-photon counting decay time apparatus has been added.

were calculated. The following pairs of K and τ were used: 10,000, 40 nsec; 200, 40 nsec; 10,000, 20 nsec; and 200, 20 nsec. Fits were over the range 0–100 nsec. To show the effect of fitting differing numbers of points, fits were done with every point (101 points total), every fifth point (21 points total), and every tenth point (11 points total). The necessary w_i, derived from Eqs. (5-11) and (5-8), are

$$w_i = 1/D(t_i) \qquad (5\text{-}12)$$

Table 5-1 summarizes the results. The weighted fits are more accurate and precise than the unweighted ones. In a single experiment the chances of obtaining a nearly correct value can increase dramatically [e.g., about three times closer for the noisy (200, 20 nsec) data set]. The discrepancies between the weighted and unweighted fits decrease for relatively noise-free data ($K = 2000$) and long lifetimes ($\tau = 40$ nsec). In these cases the w_i vary less, and the weighted fits more closely approximate the unweighted ones. Accuracy and precision fall as the number of points is reduced. This is just a signal-averaging effect. For a constant signal, the S/N is proportional to the square root of the number of points averaged.

TABLE 5-1

Effect of Weighting Factor on Exponential Fits to Semilogarithmic Plots

Assumed[a]			Calculated[b]			
			Unweighted		Weighted	
K	τ	N^c	K	τ	K	τ
2000	20	101	2020 ± 37	19.84 ± 0.23	1999 ± 14	20.04 ± 0.11
		21	2026 ± 84	19.81 ± 0.52	2002 ± 24	20.00 ± 0.21
		11	2000 ± 98	20.01 ± 0.62	2000 ± 32	20.05 ± 0.26
200	20	101	204 ± 12	19.4 ± 0.7	198 ± 4	20.5 ± 0.3
		21	207 ± 29	19.4 ± 1.5	198 ± 9	20.5 ± 0.6
		11	204 ± 28	19.8 ± 1.6	198 ± 10	20.5 ± 0.9
2000	40	101	2002 ± 15	39.91 ± 0.30	2001 ± 12	40.01 ± 0.23
		21	2004 ± 30	39.87 ± 0.62	2004 ± 21	39.94 ± 0.45
		11	1996 ± 37	40.1 ± 0.8	2000 ± 29	40.01 ± 0.59
200	40	101	201 ± 5	39.4 ± 1.0	200 ± 4	40.3 ± 0.7
		21	202 ± 10	39.3 ± 2.0	201 ± 7	40.0 ± 1.4
		11	199 ± 12	40.2 ± 2.5	200 ± 9	40.5 ± 1.9

[a] Noise is assumed to be Poisson.
[b] Results are the average and standard deviations for 100 simulations.
[c] Number of evenly spaced points used in a fit over 0–100 nsec.

Thus, using ten times the number of points improves the precision by ~3 $(10^{1/2})$; this will not be exactly true because $D(t)$ varies with t, but the correlation holds reasonably well.

We conclude from Table 5-1 that, for quality data with many points, the weighted fits are more accurate and precise than the unweighted ones, but the differences are usually inconsequential. For noisy data where the highest accuracy is required, however, proper weighting is essential. Further, for more complex functions (e.g., the sum of exponentials), proper weighting is essential to obtain the most accurate and precise results. See Section D.

Another problem arises if the data to be fit are functions of two or more measured values each with uncertainty. For example, $D(t)$ might be derived from a signal from which it was necessary to strip off a baseline with noise. How should the data be treated? In this case one uses standard formulas (Bevington, 1969; Beers, 1962) for the propagation of errors in the measured quantities to the σ's of the derived quantities.

D. NONLINEAR LEAST SQUARES

Many functions such as the sum of exponentials and Förster energy transfer data cannot be readily converted to a form linear in the parameters. For example, there is no linear form of the sum of two exponentials [Eq. (3-22) with $N = 2$], and the least squares method does not yield a closed form solution. If one applies Eqs. (5-1) and (5-4) to Eq. (3-22), nonlinear equations in the K_i and σ_i, which have no closed form solution, result. The problem is how to solve these equations for the best-fit parameters. There are many different approaches to their solution. Excellent discussions of different methods along with FORTRAN programs for implementing them are given by Bevington (1969) and Daniels (1978). We describe briefly several strategies and develop several others in more detail. All methods make initial guesses of the parameters and use an iterative process which, hopefully, improves on the guesses after each iteration.

Before discussing specific methods, we point out that caution is essential in interpreting the significance of parameters obtained by fitting. Many data sets can be fitted equally well by more than one function. Consider, for example, the two functions (Grinvald and Steinberg, 1974)

$$D_1(t) = 7500 \exp(-t/5.5) + 2500 \exp(-t/8) \tag{5-13a}$$

$$D_2(t) = 2500 \exp(-t/4.5) + 7500 \exp(-t/6.7) \tag{5-13b}$$

On a linear scale these apparently greatly different functions are indistinguishable within a pen's width over the time interval 0–50. Both functions decay from 10^4 to 4–5 at $t = 50$. Figure 5-5 shows the differences $D_1(t) - D_2(t)$.

For comparison, we have assumed that a data set was derived from Eq. (5-13a), but with Poisson SPC noise superimposed on it. The noise component shown in Fig. 5-5 represents the difference between a simulated SPC decay curve and Eq. (5-13a). Clearly, the SPC noise far exceeds the differences between $D_1(t)$ and $D_2(t)$. Indeed the sum of the squares of the differences between $D_1(t)$ and $D_2(t)$ is 1.2×10^4, whereas the sum of the squares of the single-photon counting noise is 8.5×10^4 or seven times larger. This example demonstrates that some data sets are extremely insensitive to fitting parameter variations. Grinvald and Steinberg (1974) give two expressions that are experimentally identical even though one is a sum of two exponentials and the other is the sum of three exponentials. It is, thus, possible to have a statistically acceptable fit with large errors in the fitting parameters or even an incorrect model.

We turn now to nonlinear fitting schemes. An obvious minimization approach is a "grid search" (Bevington, 1969). One makes initial guesses for the a_i and then varies a_1 by small amounts while holding the remaining a_i fixed. χ^2 is evaluated for each set of a_i. Then a_1 is varied until a minimum in χ^2 is found. Since it is unlikely that the search moved directly to the minimum, this apparent minimum is probably not the true one. See Fig. 5-1 where motion along only a_1 or a_2 does not directly approach the minimum. The search is then continued by varying a_2 while holding the remaining a_i fixed. The process of searching by varying each parameter in turn is repeated until all the parameters have been varied. This revised set of parameters yields a lower χ^2, but it is still unlikely to be the true minimum. The process is started over with a_1 and is repeated until the minimum is approached as closely as desired.

The grid search method has many attributes. It is stable; the χ^2 im-

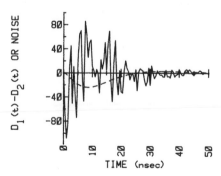

Fig. 5-5. Difference (–––) between the two sums of exponentials given by Eqs. (5-23a) and (5-23b). The solid line is the noise level that would be observed for data of the form of Eq. (5-23a) if measured on a single-photon counting instrument.

proves or becomes no worse on each iteration. The derivative information needed by many other methods is not required, and programming it is trivial. The grid search is, however, one of the slowest methods. The more efficient simplex search procedure that varies all the parameters simultaneously is described in Section D.1.

Another popular and efficient minimization procedure is the gradient search or method of steepest descent. The gradient of a function points in the direction of maximum rate of increase. The opposite direction is the path of maximum rate of decrease or the path of steepest descent. The gradient method searches from the current guess along the direction of steepest descent in χ^2 until a minimum is found. This minimum is rarely the true minimum, because the original search direction generally does not move directly toward the true minimum, although the search direction is better than the grid search. At this apparent minimum a new search direction is evaluated, and the search is resumed in the new direction. Calculation of a new search direction after each step is computationally too slow to be attractive. The gradient search is stable, reasonably efficient, and easy to program. It is an excellent search method for steep-walled surfaces as long as one is not too near the minimum. Near a minimum, however, the χ^2 surface flattens and the gradient method becomes very inefficient.

Other approaches use analytical Taylor series expansions of χ^2 or of the fitting function. Generally, these expansions are truncated after the first- or second-order terms. A normal least squares minimization is then carried out to determine the changes in each of the parameters that minimizes χ^2. If the truncated expansion is a good approximation, the corrected guesses are better than the original ones. Near the minimum, the analytical approach usually improves because of the decreasing importance of the neglected higher-order terms. Analytical approaches require derivative information, but in return generally converge faster than brute force search techniques. Analytical approaches can, however, become unstable far from the minimum where the expansion is inaccurate. Indeed, it is common to combine a simplex or gradient search for the early portions of the search and then switch to an analytical solution as the minimum is approached.

In Section 2 a widely used analytical approach based on series expansion of the function is discussed. Convergence problems are demonstrated. In Section 3 the powerful and widely used Marquardt method is described. This method automatically changes from a gradient search far from the minimum to an analytical solution near the minimum. The Marquardt method is stable and fast and is implemented with only minor modifications of the analytical method.

1. Simplex Method

The simplex method is a very valuable minimization method. This technique is basically an opportunistic empirical search through the parameter space. Unlike the simple grid search, all parameters are varied simultaneously. The simplex method is stable and easy to understand and program. It requires no derivative information. For some problems it converges nearly as rapidly as more sophisticated approaches, but for three- or four-parameter problems it is invariably slower. Also, it does not provide statistical errors in the final parameters.

Many variations of the simplex method are variations of the method of Nelder and Mead (1965) which, in turn, was an improved version of the method of Spendley et al. (1962). Our treatment follows that given by Daniels (1978) using his recommended parameter variations. We omit descriptions of the wide use of simplex optimization of experimental design (Shavers et al., 1979). There have also been attempts to enhance the simplex method by directing the search using gradient-like information (Ryan et al., 1980), but the results do not seem to justify the added complexity.

The problem is to minimize $\chi^2(a_1, a_2, \ldots, a_n)$, where $a_1, \ldots, a_n$ are the n parameters varied to minimize χ^2. A simplex is a geometric figure. In an n-dimensional space, $n + 1$ different points are needed to define the simplex. In two dimensions, the simplex is a triangle, and in three dimensions it is a tetrahedra-like figure. For $n > 3$, the simplex is not visualizable.

We now develop the general simplex minimization equations: The points in the n-dimensional space are represented by vectors. To start the simplex procedure, we need a grid of $n + 1$ guesses which, hopefully, are close to the correct values. These guesses are $\mathbf{P}_1, \mathbf{P}_2, \ldots, \mathbf{P}_n, \mathbf{P}_{n+1}$, where each $\mathbf{P}$ represents a set of guesses for the parameters. If we make initial guesses for the individual parameters a_1 through a_n, then in vector notation $\mathbf{P}_1$, our initial guess for the solution, is

$$\mathbf{P}_1 = P(a_1, a_2, \ldots, a_n, a_{n+1}) \tag{5-14}$$

$\mathbf{P}_2$ to $\mathbf{P}_{n+1}$ are different points on the simplex and are a reasonable arrangement around $\mathbf{P}_1$. A simple way of generating the remaining n $\mathbf{P}$'s from $\mathbf{P}_1$ is

$$\mathbf{P}_2 = P(1.1a_1, a_2, \ldots, a_n, a_{n+1})$$
$$\mathbf{P}_3 = P(a_1, 1.1a_2, \ldots, a_n, a_{n+1}) \tag{5-15}$$
$$\mathbf{P}_{n+1} = P(a_1, a_2, \ldots, a_n, 1.1a_{n+1})$$

As shown, all a_i must be nonzero to generate $n + 1$ independent points. One then evaluates the χ^2's at each $\mathbf{P}_i$:

$$\chi^2(\mathbf{P}_i), \quad i = 1, \ldots, n + 1$$

One then determines the $\mathbf{P}_i$ yielding the highest χ^2 (denoted by $\mathbf{P}_H$), the next highest χ^2 (denoted by $\mathbf{P}_{NH}$), and the lowest χ^2 (denoted by $\mathbf{P}_L$).

The strategy is to move the simplex through parameter space to search for the minimum χ^2. The search uses three operations: reflection, expansion, and contraction. See Fig. 5-6 for the following discussion. We consider one operation at a time.

a. Reflection

It is reasonable to assume that the best guess will be far removed from $\mathbf{P}_H$, possibly on the other side of the remaining points. Calculate the centroid $\mathbf{C}$ of all the points excluding $\mathbf{P}_H$:

$$\mathbf{C} = \frac{1}{n} \sum_{i=1, i \neq H}^{n+1} \mathbf{P}_i = \frac{1}{n} \left(\sum_{1=1}^{n+1} \mathbf{P}_i - \mathbf{P}_H \right) \tag{5-16}$$

For the triangle, $\mathbf{C}$ is midway between $\mathbf{P}_L$ and $\mathbf{P}_{NH}$ and is the center of gravity of all points excluding $\mathbf{P}_H$. Reflect $\mathbf{P}_H$ through $\mathbf{C}$ to form the reflected point $\mathbf{P}_R$ by

$$\mathbf{P}_R = (1 + \alpha)\mathbf{C} - \alpha\mathbf{P}_H \tag{5-17}$$

where α is a reflection coefficient. Nelder and Mead (1965) used $\alpha = 1$, which places $\mathbf{P}_R$ directly opposite $\mathbf{P}_H$ through $\mathbf{C}$ and the same distance from $\mathbf{C}$ as is $\mathbf{P}_H$. To avoid oscillations α is set off of 1 (e.g., 0.9985).

There are three possible results of a reflection. First, the reflection may be moderately successful:

$$\chi^2(\mathbf{P}_L) \leq \chi^2(\mathbf{P}_R) < \chi^2(\mathbf{P}_H)$$

That is, $\mathbf{P}_R$ is better than $\mathbf{P}_H$ but worse than $\mathbf{P}_L$. In this case, we form a new simplex by replacing $\mathbf{P}_H$ with $\mathbf{P}_R$. The process of reflection and testing is restarted using the new simplex.

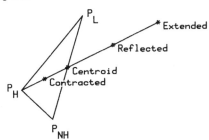

Fig. 5-6. Basic operations on a two-variable simplex.

b. Expansion

The reflection may be highly successful:

$$\chi^2(\mathbf{P}_R) < \chi^2(\mathbf{P}_L)$$

In this case we appear to be moving in an advantageous direction and expand our search further in this direction to form a new expanded point $\mathbf{P}_E$ by

$$\mathbf{P}_E = \mathbf{P}_R + (1 - \gamma)\mathbf{C} \tag{5-18}$$

where γ is the expansion coefficient. Nelder and Mead (1965) used $\gamma = 2.0$ which moves $\mathbf{P}_E$ twice as far from $\mathbf{C}$ as $\mathbf{P}_R$ and in the same direction. To avoid instabilities γ is normally slightly different than 2 (e.g., 1.95). If this expansion is successful, that is,

$$\chi^2(\mathbf{P}_E) < \chi^2(\mathbf{P}_R)$$

we replace $\mathbf{P}_H$ with $\mathbf{P}_E$. Otherwise we replace $\mathbf{P}_H$ with $\mathbf{P}_R$. The process of reflecting and testing is then restarted.

c. Contraction

If the initial reflection fails, that is,

$$\chi^2(\mathbf{P}_R) \geq \chi^2(\mathbf{P}_H)$$

then the minimum may be on the same side of $\mathbf{C}$ as is $\mathbf{P}_H$. Therefore, we generate a contracted point $\mathbf{P}_C$ given by

$$\mathbf{P}_C = (1 - \beta)\mathbf{C} + \beta\mathbf{P}_H \tag{5-19}$$

For $\beta = 0.5$, $\mathbf{P}_C$ is halfway along the line connecting $\mathbf{P}_H$ and $\mathbf{C}$. Again, to avoid instabilities, $\beta = 0.4985$. If contraction is successful,

$$\chi^2(\mathbf{P}_C) < \chi^2(\mathbf{P}_H)$$

then $\mathbf{P}_H$ is replaced by $\mathbf{P}_C$ and the reflection and testing procedure is restarted.

Contraction rarely fails, except as one closely approaches the minimum. If it does, however, we redefine the simplex and try again. Scaling has been suggested to accomplish this (Daniels, 1978). Every point $\mathbf{P}_i$ is replaced by

$$\mathbf{P}_i + k(\mathbf{P}_L - \mathbf{P}_i), \qquad i = 1, \ldots, n + 1 \tag{5-20}$$

where k is the scaling constant. $\mathbf{P}_L$ is left intact, but all other points change as follows: If $k = 0.5$, every point moves toward $\mathbf{P}_L$ and halves the distance. If $k = -2$, every point moves away from $\mathbf{P}_L$ and doubles its distance. If $k = 2$, every point moves toward $\mathbf{P}_L$ but overshoots and ends up as far away as originally but on the opposite side. In programs that run unattended, k is set to 0.5 (Nelder and Mead, 1965). In an interactive program, k is frequently an operator input.

In an overview, as the solution develops, the simplex tumbles through parameter space. When it finds an advantageous pathway, it stretches along it. When it finds a wall it backs away. Finally, as it homes in on the final parameter set, it contracts.

To show the utility of the simplex method we use it to fit a 100-point simulated excimer kinetics decay curve

$$D(t) = 18{,}692[\exp(-t/35) - \exp(-t/7)] \tag{5-21}$$

over the 0–99-nsec range. The preexponential factor gives a peak of 10^4 counts. The Poisson noise characteristic of an SPC instrument was added before fitting. The simulated decay, $D(t_i)$ ($i = 0$ to 99), was reduced using a three-parameter simplex fit with $F(t_i)$ given by

$$F(t_i) = K_1[\exp(-t_i/\tau_1) - \exp(-t_i/\tau_2)] \tag{5-22}$$

A general purpose BASIC program SIMPLEX used to carry out the fit is given in Appendix E. The program was run on an HP-85A desk-top microcomputer. Table 5-2 shows a partial listing of the search. Convergence was decreed when the fractional change in χ^2 between iterations

TABLE 5-2

Convergence of a Simplex Fit to Excimer Kinetics

Iteration	K	τ_1	τ_2	$\chi^2 \times 10^{-5}$
0	13,000	40.00	5.000	963.7
1	13,866	42.66	4.501	265.2
2	14,442	44.44	4.667	198.8
3	14,435	41.31	4.768	197.5
4	14,285	44.20	5.013	176.5
5	15,019	42.49	4.785	161.3
10	15,319	40.16	5.149	105.4
15	15,930	40.08	5.618	64.7
20	17,028	38.15	6.380	28.7
25	17,772	36.70	6.531	14.6
30	18,068	35.81	6.773	8.6
35	18,389	35.417	6.893	6.1
40	19,040	34.52	7.112	4.61264
45	18,977	34.64	7.098	4.58708
46	18,949	34.67	7.082	4.58619
47	18,976	34.63	7.094	4.58513
48	18,960	34.65	7.088	4.58509

was $<10^{-5}$. An unweighted fit required 48 iterations and 35 min using Eq. (5-22) in the χ^2 evaluation. Use of the fast iterative exponential evaluation (Section G) would probably have better than halved the time. As in many minimizations, the search does not always proceed in what seems like a direct route. The first iteration produces the largest change in χ^2, yet the estimates of τ_1 and τ_2 are poorer.

The most time-consuming part of the fit was the χ^2 calculation. To accelerate convergence, the data set was reduced by originally calculating χ^2 with only every tenth point. Once the fractional change in χ^2, $|\Delta\chi^2/\chi^2|$, between iterations was $<10^{-3}$, every point was used in calculating χ^2. Convergence was considered complete when $|\Delta\chi^2/\chi^2|<10^{-5}$. The convergence criteria were varied because it does little good to refine a partial data set carefully when the inclusion of more data will cause large parameter changes. In this example, after 20 iterations (~3 min), every point was used in the calculation. Forty iterations (~17 min) were required, thereby halving the calculation time. Thus, if reasonable initial parameter guesses are not known, the strategy of using a relatively limited number of points to approach the correct solution is highly advantageous. This approach is suitable for any minimization method, but the number of points fit should always be several times larger than the number of parameters.

We routinely use SIMPLEX for solving nonlinearizeable functions. For example, we have fit the quenching of tris(2,2'-bipyridine)ruthenium(II) by $HgCl_2$ with increasing $[Cl^-]$. As $[Cl^-]$ increases, $HgCl_3^-$ and $HgCl_4^{2-}$ form, each with its own association and quenching constants. Using SIMPLEX we have evaluated the quenching constants for $HgCl_3^-$ and $HgCl_4^{2-}$ and association constants for each species (Hauenstein et al., 1983b).

SIMPLEX is adaptable to any nonlinear system. When used on a microcomputer, however, it is likely to be too slow for $n > 4$, although we occasionally use it with $n = 6$.

2. Analytical Solution

We linearize the nonlinear equations formed from Eq. (5-4) by expanding $F(x, a_1, \ldots, a_p)$ in a Taylor series around some initial guess (Bevington, 1969; Rogers, 1962; Johnson, 1980; Scarborough, 1962). The initial guesses for a_1 to a_p are q_1 to q_p; then the a_i are

$$a_1 = q_1 + \delta a_1$$
$$\vdots \qquad\qquad (5\text{-}23)$$
$$a_p = q_p + \delta a_p$$

The unknown δa_j are corrections to q_j and are assumed to be small enough to expand F in a Taylor series in the δa_j and truncate after the first-order terms:

$$F(x_i, a_1, a_2, \ldots, a_p) = F(x_i, q_1, \ldots, q_p) + \sum_{j=1}^{p} d_{ji} \, \delta a_j \quad (5\text{-}24)$$

$$d_{ji} = (\partial F/\partial a_j)_i = [\partial F(x_i, a_1, \ldots, a_p)/\partial a_j]_{a_k, k \neq j}$$

The a_i are given by Eq. (5-23). Because of the truncation, Eq. (5-24) works best for q_j near the correct values.

After making initial guesses, everything in Eq. (5-24) is known except the δa_j. We can now substitute Eq. (5-24) into Eqs. (5-1) and (5-4) and apply a normal linear least squares solution to yield the a_j. Then using Eq. (5-23) we generate new a_j. If Eq. (5-24) were exact, the new a_j would be correct. In general, however, the approximation in Eq. (5-24) means that the new a_j are only better approximations than the original guesses. These new a_j are then used as new guesses, q_j, and the entire procedure is repeated to generate a new set of a_j and new a_j. For well-behaved functions, each iteration reduces χ^2. Near the minimum, convergence can become exceedingly rapid. Note that we are solving for corrections to these parameters rather than for the parameters themselves.

To demonstrate this technique we will fit a simple exponential decay using a nonlinear approach:

$$y = F(t) = K \exp(-t/\tau) = a_1 \exp(-t/a_2) \quad (5\text{-}25)$$

For consistency with our equations we use a_1 in place of K and a_2 in place of τ. The initial guesses for a_1 and a_2 are q_1 and q_2. From Eq. (5-24)

$$F(t_i, q_1 + \delta a_1, q_2 + \delta a_2) = q_1 \exp(-t_i/q_2) + d_{1i} \, \delta a_1 + d_{2i} \, \delta a_2$$

$$d_{1i} = (\partial F/\partial a_1)_i = \exp(-t_i/a_2) \quad (5\text{-}26)$$

$$d_{2i} = (\partial F/\partial a_2)_i = (a_1 t_i/a_2^2) \exp(-t_i/a_2)$$

For N data points, there are N of both d_{1i} and d_{2i}.

We consider briefly the quality of the approximation of Eq. (5-24). It can readily be shown that the approximation of Eq. (5-26) is reasonably good ($<40\%$ error) for up to 20% errors in a_1 and a_2 and for t/a_2 from 0 to 5. Deviations become much larger for poorer q's, and convergence problems might arise for less accurate initial guesses.

To the level of approximation of Eq. (5-26), χ^2 is given by Eq. (5-1a) with

$$R_i = y_i - F(x_i, q_1, q_2) - d_{1i}\delta a_1 - d_{2i}\delta a_2 \quad (5\text{-}27)$$

As in the linear case, we minimize χ^2 with respect to both δa_1 and δa_2 by

$$(\partial \chi^2 / \partial \delta a_1), = (\partial \chi^2 / \partial \delta a_2) = 0 \qquad (5\text{-}28)$$

which yields the two linear equations in δa_1 and δa_2:

$$b_{11}\delta a_1 + b_{12}\delta a_2 = c_1 \qquad (5\text{-}29a)$$

$$b_{21}\delta a_1 + b_{22}\delta a_2 = c_2 \qquad (5\text{-}29b)$$

$$b_{kl} = \Sigma\, w_i d_{ki} d_{li} \qquad (5\text{-}29c)$$

$$c_k = \Sigma\, w_i d_{ki} R_i \qquad (5\text{-}29d)$$

This pair of equations has the solutions

$$a_1 = (c_1 b_{22} - c_2 b_{12})/\Delta$$

$$a_2 = (c_2 b_{11} - c_1 b_{21})/\Delta \qquad (5\text{-}30)$$

$$\Delta = b_{11} b_{22} - (b_{12})^2$$

The summations are over all points used in the fit. We now apply Eqs. (5-29) and (5-30) to minimization of the SPC data similar to that of Fig. 5-4. We make initial guesses q_1 and q_2 for a_1 and a_2 and calculate δa_1 and δa_2. Using Eq. (5-23), we calculate new and, hopefully, improved a_1 and a_2. We can verify that the χ^2 for the new a_1 and a_2 is an improvement. The new a_1 and a_2 are our guesses for the next iteration, and we repeat this process. Table 5-3 shows the results for fitting a decay curve; the parameters, before adding Poisson noise, were $a_1 = 10{,}000$ and $a_2 = 40$ nsec. The fit was to 100 points over the 0–99-nsec range.

The first entry is for initial guesses of $q_1 = 9500$ and $q_2 = 55$ nsec. Convergence was assumed when the fractional change in χ^2 between succeeding iterations was $<10^{-5}$, which required four iterations; the fourth iteration is omitted, since the discrepancies are too small to show up. After each iteration, the χ^2 was reduced. Similar results were obtained for any combination of guesses for q_1 of 7500 to 12,500 and q_2 of 10–60 nsec. Typically four to five and occasionally six iterations were required. Thus, even many unrealistic initial guesses yield rapid convergence. The second set of initial guesses in Table 5-3, however, show that some poorer guesses can give convergence, but the χ^2 may actually increase during part of the search. Finally, the solution is quite forgiving even if only one parameter is known accurately. For example, if $q_2 = 40$, then q_1 for 100–

TABLE 5-3

Convergence of Nonlinear Least Squares Fits for Different Initial Guesses[a]

Iteration	Simple analytical method			Marquardt method		
	K	τ	χ^2	K	τ	χ^2
0	9500.00	55.00	30983	9500.00	55.00	30983
1	9775.01	38.86	923.55	9767.93	38.91	902.57
2	10007.28	39.98	97.81	10007.50	38.98	97.81
3	10009.56	39.95	97.71	10009.56	39.95	97.71
0	3500.00	60.00	111841	3500.00	60.00	111841
1	9657.66	0.75	355761	7870.26	33.32	40492[b]
2	9334.20	3.21	321551	9886.76	41.80	585.0
3	5297.65	13.54	247456	10002.40	39.94	98.11
4	6194.83	58.86	25375	10009.56	39.95	97.71
5	9684.65	27.40	43922			
6	9475.84	40.79	700.71			
7	10008.26	39.90	98.29			
8	10009.56	39.95	97.71			
0	3000.00	60.00	138769	3000.00	60.00	138769
1	9655.66	−9.13	1.3E15	7763.76	30.76	56050[b]
2	1.03	−9.13	1.4E6	9972.06	42.71	1020.08
3	1.03	−9.98	2.5E5	9994.97	39.89	100.31
4	2.44	−12.37	5.5E5[c]	10009.55	39.95	97.71

[a] All simulations used the single-photon counting single-exponential decay with $K = 10,000$ and $\tau = 40$. The fitting region was for 100 points of t (0–99). The initial guesses are iteration 0.

[b] The initially calculated χ^2 was larger than the value calculated using the initial guesses. Lambda was thus increased by a factor of 10, and recomputation of the δa_j was repeated. The indicated decreased chi-square was then obtained.

[c] The χ^2 continued to diverge for later iterations.

10^5 give rapid convergence, as do initial values of q_1 of 10,000 and q_2 in the range 5–130 nsec.

The third set of initial guesses shows that too poor a starting point can yield a divergent solution. Further, a divergent solution can occur even if only one of the parameters is inaccurate (e.g., $a_1 = 10,000$, $a_2 = 135$). Thus, in this analytical solution, good initial guesses are important. This is especially true, as we shall show, if more than two parameters are being fit.

This nonlinear least squares approach is easily generalized to more parameters. Using Eq. (5-23) with p parameters and carrying through the setting of the p partial derivatives with respect to the δa_j equal to zero yields p equation in p a_i unknowns as follows

$$b_{11}\delta a_1 + b_{12}\delta a_2 + \cdots + b_{1p}\delta a_p = c_1$$
$$b_{21}\delta a_1 + b_{22}\delta a_2 + \cdots + b_{2p}\delta a_p = c_2$$
$$\vdots \qquad\qquad (5\text{-}31a)$$
$$b_{p1}\delta a_1 + b_{p2}\delta a_2 + \cdots + b_{pp}\delta a_p = c_p$$

$$b_{kl} = \Sigma\, w_i d_{ki} d_{li} \qquad (5\text{-}31b)$$

$$c_k = \Sigma\, w_i d_{ki} R_i \qquad (5\text{-}31c)$$

Since the coefficients are symmetric (i.e., $b_{kl} = b_{lk}$), it is necessary only to calculate half the off-diagonal elements. As before, one solves for the δa_j and then the new a_j. This process is repeated until convergence is obtained.

A reasonable convergence criteria is that χ^2's for succeeding iterations agree to within some error bounds such as 1 part in 10^5. Alternatively, all $|\delta a_j/q_j|$ terms should fall below a certain level (e.g., 10^{-5}). Some programs demand that both criteria be satisfied. In our calculations convergence is decreed when either of these conditions is satisfied. Two points bear mentioning. The larger the number of unknowns, the more sensitive the solution is to the poor initial guesses, and as data quality decreases, the number of parameters that can be accurately fit falls.

We have described four methods for fitting a single exponential decay: weighted and unweighted linear least squares fits to the semilogarithmic plots and weighted and unweighted nonlinear least squares fits. Which

TABLE 5-4

Comparison of Different Evaluation Methods for a Single-Exponential Decay[a]

Linear fit to semilogarithmic plot[b]		Nonlinear least squares fit[b]	
Unweighted	Weighted	Unweighted	Weighted
10010 ± 51	9993 ± 28	10000 ± 34	9999 ± 28
19.98 ± 0.06	20.01 ± 0.05	19.99 ± 0.09	19.98 ± 0.05

[a] The decay was of the form $D(t) = 10^4 \exp(-t/20)$ with fitting over the interval 0–99. Single-photon counting noise was added before data reduction.

[b] Each entry is the result of 20 simulations. The mean and standard deviation are given.

method is best? For a trial case we select $a_1 = 10^4$ and $a_2 = 20$ nsec with a fit over the range 0–99 nsec and then introduce noise with SPC statistics. These conditions are moderately demanding; the wide range in $D(t)$'s causes the w_i to vary substantially.

Table 5-4 summarizes the results. Twenty different decays were generated. The mean calculated a_i and their standard deviations are given for the different data reduction approaches. The two weighted approaches are equivalent and clearly superior in both the semilogarithmic and nonlinear fits. Therefore, for a single exponential fit, a simple weighted linear least squares fit to the semilogarithmic plot is the simplest and fastest data reduction method. From a practical standpoint, however, all four approaches are likely to yield acceptable accuracy and precision, especially for smaller variations in $D(t)$ where the w_i vary less.

3. Marquardt Method

Marquardt (1963) developed a method that exhibited a gradient like search direction when far from the minimum and then moved smoothly into the analytical method near the minimum. His method is easily adapted to the analytical solution of Section 2. Marquardt left the off-diagonal elements of Eq. (5-31a) unchanged but redefined the diagonal elements as follows:

$$b_{ii} = b_{ii}(1 + \lambda), \qquad i = 1, \ldots, p \qquad (5\text{-}32)$$

If $\lambda = 0$, we have the analytical solution of Section 2. If λ is large, the off-diagonal elements b_{ij} $(i = j)$ become insignificant compared to the diagonal elements. The search direction is then along the path of steepest descent or the gradient method (Bevington, 1969).

The Marquardt method adjusts λ to ensure that after each iteration χ^2 decreases; λ is reduced at each iteration as long as χ^2 decreases. If the solution causes χ^2 to increase, however, λ is increased. In this manner, failure of the analytical-like solution causes λ to increase, which makes the solution more steepest-descent-like until χ^2 is reduced. As the minimum is approached, however, the analytical solution usually becomes more accurate and λ approaches zero.

The rules for carrying out the Marquardt minimization (Bevington, 1969) are as follows:

(1)   Make initial guesses for the a_i equal to $q_1, \ldots, q_p$. Calculate an initial χ^2 for these values to yield χ_i^2.

(2)   Set $\lambda = 0.001$ initially and generate the b_{ii} using Eq. (5-32a).

(3)   Solve for the δa_j by Eq. (5-31).

(4) Solve for the trial a_i by Eq. (5-23) and evaluate χ^2 using these trial values to yield χ_T^2.

(5) If $\chi_T^2 < \chi_I^2$, then the trial solution was acceptable. Reduce λ to 0.1λ, replace the old q_i by the current a_i and χ_I^2 by χ_T^2 and go to step 3.

(6) If $\chi_T^2 \geq \chi_I^2$, the trial solution was unacceptable. Increase λ to 10λ and return to step 3.

Thus, the Marquardt approach increases λ as required to ensure that χ^2 decreases on each iteration and reduces λ to zero as convergence is approached. It may be necessary in a single iteration to vary λ and solve for the δa_j several times before χ^2 decreases. Since the b_{ij} remain unchanged, the only extra computations are solving for the δa_j and the evaluation of χ_T^2. For large data sets evaluation of the b_{ij} is the most time-consuming part. Therefore, the Marquardt method always gives a monotonic reduction in χ^2 and, frequently, increased speed and probability of convergence.

Table 5-3 demonstrates the results of the Marquardt algorithm applied to the same SPC data of Section 2. These results demonstrate several general points. For good initial guesses, the analytical and Marquardt methods are virtually identical. But for poorer guesses, the Marquardt approach converged monotonically and faster than the analytical solution and solved cases that diverged with the analytical method.

We turn now to the much more formidable problem of fitting a sum of two exponentials [Eq. (3-22) with $N = 2$]. This is a much more complex problem, since there are four parameters to fit. We have carried out a large number of simulations with different parameters and initial guesses to demonstrate the performance of the nonlinear least squares fitting.

For the most exhaustive study, we have selected $\tau_1 = 35$ nsec and $\tau_2 = 7$ nsec with K_1 from 500 to 9000 using $K_1 + K_2 = 10,000$. Also, reduction was carried out with $\tau_1 = 35$ nsec, $\tau_2 = 15$ nsec, and $K_1 = K_2 = 5000$. In all cases we used synthetic SPC data. Data reductions were carried out by the Marquardt nonlinear least squares fitting using weighted and unweighted fits. In all cases 100 data points in the range 0–99 nsec were used in the fits. The program EXPFIT (Appendix E) was used.

The 35- and 7-nsec lifetimes are reasonably well separated, and reduction is reasonably straightforward. The 35- and 15-nsec lifetimes are sufficiently close that the fitting problem is more severe.

Before discussing the results, we consider convergence problems that are significantly more serious here than with the single-exponential case. As an example of how even the Marquardt algorithm can suffer from slow convergence, we generated a 100-point noise-free double-exponential decay curve with $K_1 = 9000$, $\tau_1 = 35$ nsec, $K_2 = 1000$, and $\tau_2 = 7$ nsec ($t = 0$ to 99 nsec). This data set was then reduced using an SPC-weighted fit with different initial guesses. Regardless of the weighting scheme or initial guesses, the solution should converge to the initial parameters. For initial guesses of $K_1 = 10,500$, $\tau_1 = 36$ nsec, $K_2 = 0$, and $\tau_2 = 0.5$ nsec, χ^2 reduces to 10^{-9} in six iterations. When, however, we make the same guesses except for $\tau_2 = 100$ nsec, the search through parameter space becomes extremely tortuous and inefficient. After 30 iterations χ^2 was reduced to only 55.6. At 50 iterations $\chi^2 = 44.6$ with $K_1 = 5490$, $\tau_1 = 26.0$, $K_2 = 4213$, and $\tau_2 = 44.6$. At 60 iterations $\chi^2 = 34.5$ with $K_1 = 3005$, $\tau_1 = 21.2$, $K_2 = 6741$, and $\tau_2 = 34.5$. At this point, however, convergence became very rapid, and $\chi^2 = 10^{-17}$ at 69 iterations with $K_1 = 1000$, $\tau_1 = 7$, $K_2 = 9000$, and $\tau_2 = 35$. Clearly, the reason for the slow convergence was the need to convert the high amplitude, short lifetime guess into the low amplitude, short lifetime component; apparently there is no direct route for this conversion. Of course, in this case visual inspection of the early fits would show the unsuitability of such a long guess for τ_2.

The complexity of the χ^2 surfaces for two exponential fits is beautifully demonstrated by the contour maps for a sum of two-exponential data presented by Hinde et al. (1977).

These seemingly simple cases demonstrate the sensitivity of the search to the qualities of the initial guesses. Thus, always make the initial guess as accurate as possible. If convergence problems arise, change the guesses. In a two-exponential fit, change the longer-lived guess to the shorter one. Particularly valuable for making or refining guesses is visual comparison of calculated and observed data. Alternatively, initially use a simplex fit with a coarse grid to locate reasonable starting parameters.

Table 5-5 summarizes the results of the double-exponential fits. For every entry 20 simulations were carried out, and the mean and standard deviation are given for each parameter. Several important points emerge. For all cases, the calculated K's and τ's agree with the expected ones within statistical uncertainty (typically <1 standard deviation of the mean $= \sigma/20^{1/2}$). In all cases, except for the entries with large K's associated with long τ or for the close lifetime sets, weighting greatly improves the results. The case of $K_1 = 500$ and $\tau_1 = 35$ nsec is especially dramatic. The standard deviations for K_1 and for τ are over two times greater for the unweighted versus the weighted data fits. This disparity arises because the unweighted fit overweights the noisy low-amplitude tail from the 35-nsec component. In contrast, for small K's in the fast component, weighting has little effect; the short-component contribution is predominately at short times where the amplitude is relatively constant and noise-free.

It is clear that the correct weighting approach should be carried out whenever possible. At least with SPC data this requires little or no additional effort and either gives improved accuracy or is equivalent to unweighted fits.

TABLE 5-5

Results of Double-Exponential Fits by Nonlinear Least Squares

Actual values[a]		Method[b]	Calculated values[c]			
K_1	K_2		K_1	τ_1	K_2	τ_2
500	9500	U	513 ± 62	34.8 ± 2.6	9501 ± 72	6.94 ± 0.10
		W	512 ± 28	34.3 ± 0.9	9501 ± 58	6.87 ± 0.08
1000	9000	U	1012 ± 65	34.9 ± 1.4	9004 ± 76	6.97 ± 0.11
		W	1013 ± 33	34.7 ± 0.6	9002 ± 61	6.97 ± 0.09
2000	8000	U	2011 ± 72	34.9 ± 0.8	8005 ± 83	6.96 ± 0.13
		W	2013 ± 41	34.8 ± 0.4	8002 ± 65	6.96 ± 0.12
5000	5000	U	5010 ± 89	35.0 ± 0.4	5007 ± 102	6.9 ± 0.2
		W	5013 ± 56	34.9 ± 0.2	5004 ± 76	7.0 ± 0.2
8000	2000	U	8006 ± 103	35.0 ± 0.3	2014 ± 118	6.8 ± 0.6
		W	8007 ± 72	35.0 ± 0.2	2012 ± 87	6.8 ± 0.7
9000	1000	U	8996 ± 110	35.0 ± 0.3	1026 ± 124	6.7 ± 1.4
		W	8987 ± 111	35.0 ± 0.2	1031 ± 107	6.9 ± 1.7
5000	5000[d]	U	5059 ± 771	34.8 ± 2.1	4854 ± 731	14.6 ± 1.5
		W	5090 ± 633	34.9 ± 1.8	4917 ± 599	14.7 ± 1.2

[a] Unless otherwise indicated, $\tau_1 = 35$ nsec and $\tau_2 = 7$ nsec.
[b] U, Unweighted; W, single-photon counting weights.
[c] Fits were over the entire range of 0–99 nsec.
[d] $\tau_1 = 35$ nsec, $\tau_2 = 15$ nsec.

Table 5-5 has other interesting ramifications. As expected, reducing a component's contribution decreases the accuracy with which its parameters can be measured. Also, as the lifetimes approach each other, the accuracy suffers enormously. Compare the $K_1 = K_2 = 5000$ cases for the two lifetime sets. Also, compare nonlinear least squares fitting to the baseline stripping method used in Section G. Comparison of Tables 3-1 and 5-5 shows that the baseline stripping procedure using unweighted fits is generally inferior in accuracy and precision to the method of unweighted nonlinear least squares, especially for large K_1. For a low-amplitude slow component the weighted nonlinear least squares fit is clearly superior with up to two times smaller standard deviations. Thus, for accurate work, nonlinear least squares fitting is always to be preferred over baseline stripping.

E. JUDGING THE FIT

Having generated the best-fit parameters how does one judge the quality of the fit? Is it statistically acceptable? Mathematical and/or visual methods are frequently used to test the fit. We discuss two widely used visual methods, the residual and autocorrelation plots. We also describe the standard mathematical chi-square test.

From the best fit parameters, one can calculate the expected curve. Even for the correct model the theoretical and observed decays will differ because of noise. To visualize these discrepancies, one makes plots of the simple residuals R_i or weighted residuals R_{wi} versus time:

$$R_i = F(t_i) - D(t_i) \tag{5-33a}$$

$$R_{wi} = w_i R_i = R_i/\sigma_i^2 \tag{5-33b}$$

where $F(t)$ and $D(t)$ are the calculated best fit and observed decay data, respectively. The w_i and σ_i are the normal weighting factors and standard deviations of the ith point. A good model should yield randomly distributed residual plots.

To demonstrate the use of residual plots we carried out a weighted nonlinear least squares fit of a simulated single-photon counting exponential decay. The noise-free decay was $10^5 \exp(-t/12 \text{ nsec})$, and fitting was over the 0–100-nsec range. The residual plot is given in Fig. 5-7a. Since the model is a good one, the R_i are randomly distributed.

Figure 5-7a shows a problem with the simple residual plot. The larger $D(t)$, the larger the R's because the noise is given by $D(t_i)^{1/2}$. In Fig. 5-7a it is easy to judge the fit at early times where $D(t)$ is large, but hard at the end region where $D(t)$ is small. A better way to view the fit is with the weighted residuals plot. For single-photon counting data, σ_i^2 is $D(t)$, which yields

$$R_{wi} = [F(t_i) - D(t_i)]/D(t_i) \tag{5-34}$$

Strictly speaking $\sigma_i^2 = F(t_i)$; however, for a good model the differences between $F(t)$ and $D(t)$ will be inconsequential and Eq. (5-34) holds. Figure

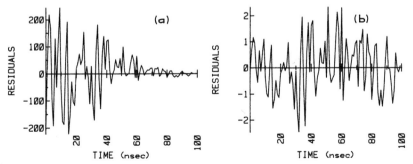

Fig. 5-7. Comparison of unweighted (a) and weighted (b) residual plots for a weighted least squares fit to a single-exponential decay. The decay was of the form $10^5 \exp(-t/12)$ and exhibited Poisson statistics.

5-7 b shows the R_{wi} for the same data as Fig. 5-7a. Clearly, visualization of the low-amplitude region is much improved.

We consider a more complex example to show further the differences between weighted and unweighted residual plots and to demonstrate the utility of these plots in detecting a model failure. We use a weighted nonlinear least squares fit of a single-exponential model to SPC data. In this case, however, the actual data are the sum of two exponentials with $K_1 = 10^5$, $\tau_1 = 12$, and $\tau_2 = 40$. Here K_2 was given values of 100, 200, and 500. This is equivalent to adding small amounts of a long-lived component to the example of Fig. 5-7a. Figure 5-8a and b shows the unweighted and weighted residual plots for the fits with $K_2 = 100$. Figure 5-8c and d shows similar data for $K_2 = 200$, and Fig. 5-8e and f is for $K_2 = 500$. Close visual

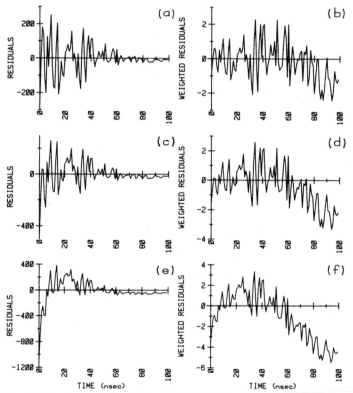

Fig. 5-8. Comparison of unweighted (a, c, e) and weighted (b, d, f) residual plots for a single-exponential fit to a sum of two exponential decays: $D(t) = K_1 \exp(-t/\tau_1) + K_2 \exp(-t/\tau_2)$. Here $K_1 = 100,000$; $\tau_1 = 12$; $\tau_2 = 40$. For (a) and (b), $K_2 = 100$. For (c) and (d), $K_2 = 200$. For (e) and (f), $K_2 = 500$. Weighted nonlinear least squares were used in all fits. If an unweighted fit is used, the unweighted residual plots of (c) and (e) look quite different.

inspection of the unweighted residual plots clearly shows failure of the unweighted single-exponential model for $K_2 = 200$ and 500; however, the true extent of this failure is much better demonstrated by the weighted residuals. For $K_2 = 100$ both the unweighted and weighted residuals suggest a failure of the model, especially at long times, but the discrepancies are not large enough to risk a professional reputation.

We stress that this type of data examination should be used with insight and caution, as it is sensitive to trivial artifacts. In our double-exponential case with $K_2 = 200$, the model failure is rather extreme, yet the emission intensity contributed by the long-lived component is only 0.7% of the total. Thus, a long-lived impurity with the same emission spectrum and photon yield as the major component could cause the observed deviation even if present as only an 0.7% impurity. Higher-yield impurities would produce the same artifact at still lower concentrations. Such levels of sample cleanliness are frequently impossible to achieve. Further, interpretations do not always require better data.

The autocorrelation function is also widely used to express the deviations (Grinvald and Steinberg, 1974):

$$C(t_j) = \left(\frac{1}{m} \sum_{i=1}^{m} R_i R_{i+j}\right) \Big/ \left(\frac{1}{N} \sum_{i=1}^{N} R_i^2\right) \tag{5-35}$$

$C(t_j)$ should be randomly distributed for random noise. We give an example in Chapter 10. For SPC instruments working near the theoretical limit, the autocorrelation function contains little or no information beyond that of the weighted residual plot. Other instruments such as sampling devices and less finely tuned SPC instruments can have systematic distortions. The autocorrelation function is then nonrandom and quite characteristic of the instrument. Indeed, Grinvald (1976) has pointed out that one of the advantages of standards is that they permit one to characterize an instrument's autocorrelation function accurately. Then, fits that deviate from the expected autocorrelation indicate an inadequate model. In contrast, such an instrument will not yield a randomly distributed residual plot.

A mathematical method of assessing the adequacy of the fit is an ordinary runs test that judges the randomness of the residuals with regard to sign. If the data are from an SPC instrument, the R_{wi} should exhibit a certain number of sign changes (see Chapter 11). Significant deviations indicate model or instrument failure. For a data set spanning a wide amplitude range, this test is extremely demanding because of instrumental and sample limitations.

The most common mathematical test is the chi-square test. One calculates the χ^2 of Eq. (5-1), or the reduced chi-square χ_r^2, and compares it with the expected value:

$$\chi_r^2 = \chi^2/v, \qquad v = N - P - 1 \tag{5-36}$$

where v is the number of degrees of freedom in the fitting of N points with P parameters. Intuitively, if the fitting function is correct, R_i/σ_i should tend toward an average absolute value of unity. Thus, for a relatively large number of points, χ^2 is about N or χ_r^2 is about unity.

We discuss how to interpret the χ^2 or χ_r^2 from an experiment. Table 5-6 is a very much abbreviated table of χ_r^2 and is read as follows: For the indicated number of degrees of freedom each entry corresponds to the χ_r^2 for which there is a probability p of measuring a χ_r^2 greater than, or equal to, the indicated value. Thus, for $p = 0.10$ and $v = 100$, $\chi_r^2 = 1.19$. This means that for an experiment with 100 degrees of freedom there is a 10% chance of measuring a $\chi_r^2 \geq 1.19$. Extensive tables are available (Bevington, 1969), but the current table is adequate for our needs, since the χ_r^2 distribution is not very sensitive to v for large values of v (>90), especially for $p < 0.95$.

Table 5-6 is used as follows: Evaluate χ_r^2 and determine the probability of observing this χ_r^2. If this probability is highly unlikely, the fit is suspect. Too small a χ_r^2 is just as unacceptable as one that is too large. Usually too small a χ_r^2 results if too many parameters are used or the σ_i are improperly evaluated.

We apply the χ^2 test to the data of Fig. 5-7 and 5-8. Since 100 points were fit to a single exponential model, $v = 97$ ($N = 100$ and $P = 2$). For the single exponential data of Fig. 5-7, $\chi_r^2 = 1.01$. From Table 5-6 we see that the probability of observing a χ_r^2 greater than this is only slightly less than 50%; thus, the model appears to be a good one. For Fig. 5-8b, $K_2 = 100$ and $\chi_r^2 = 1.26$, which corresponds to a probability slightly less than 5%.

TABLE 5-6

Distribution of Reduced Chi-squares[a,b]

p v	0.99	0.95	0.90	0.60	0.50	0.40	0.10	0.05	0.01	0.001
50	0.59	0.70	0.75	0.94	0.99	1.04	1.26	1.35	1.52	1.73
90	0.69	0.77	0.81	0.96	0.99	1.03	1.20	1.26	1.38	1.53
100	0.70	0.78	0.82	0.96	0.99	1.03	1.19	1.24	1.36	1.45
200	0.78	0.84	0.87	0.97	1.00	1.02	1.13	1.17	1.25	1.34

[a] The indicated values are the probability p of exceeding a reduced chi-square χ_r^2 for v degrees of freedom.

[b] Adapted from Bevington (1969). Copyright © 1969 by McGraw-Hill. Used with permission of McGraw-Hill Book Company.

This result makes the model suspect but, as with the visual test, does not disprove it; after all, about 1 in 20 experiments would yield a χ_r^2 greater than this value even if the model were correct. There is no question, however, about the $K_2 = 200$ and 500 cases (Fig. 5-8d and f) where the χ_r^2 are 2.06 and 6.52, respectively. In both cases there is much less than a 0.1% chance of observing the χ_r^2, and the single-exponential model fails.

For single-photon-counting data, $1/\sigma_i^2$ is just $1/F(t_i)$ or, more commonly, $1/D(t_i)$. For other instruments rigorous knowledge of σ_i is not known and χ_r^2 must be approximated. For example, once an instrument is widely used, the expected noise level becomes known and the necessary weighting factors can be estimated.

F. ERROR ESTIMATION

Ideally, the best way to estimate the errors in an experiment is to repeat the experiment several times and directly calculate the standard deviations. With complicated or time-consuming experiments, this approach is unacceptable. One can, however, estimate the uncertainty in the least squares best-fit parameters directly from a single experiment. The following expressions come from Bevington (1969).

If one considers the analytical approach of Section D.2. one can express the general normal equations, Eq. (5-31), in matrix formulation as

$$\mathbf{B}\,\delta\mathbf{a} = \mathbf{C} \tag{5-37}$$

$$\mathbf{B} = \begin{pmatrix} b_{11} & b_{12} & \cdots & b_{ip} \\ \vdots & \vdots & & \vdots \\ b_{p1} & b_{p2} & \cdots & b_{pp} \end{pmatrix}, \qquad \delta\mathbf{a} = \begin{pmatrix} a_1 \\ \vdots \\ a_p \end{pmatrix}, \qquad \mathbf{C} = \begin{pmatrix} c_1 \\ \vdots \\ c_p \end{pmatrix}$$

To solve for the δa_j, one multiplies both sides by the inverse matrix of $\mathbf{B}$, $\mathbf{B}^{-1}$:

$$\mathbf{B}^{-1}\mathbf{B}\,\delta\mathbf{a} = \mathbf{B}^{-1}\,\mathbf{C} \tag{5-38a}$$

$$\delta\mathbf{a}_j = \mathbf{B}^{-1}\,\mathbf{C} \tag{5-38b}$$

$\mathbf{B}^{-1}$ is the error matrix, and the uncertainties in the final a_j, σ_{a_j}, are (Bevington, 1969):

$$\sigma_{a_j} = (B_{jj}^{-1})^{1/2} \tag{5-39}$$

or the standard deviation in a_j is the square root of the jth diagonal element of $\mathbf{B}^{-1}$ (denoted by B_{jj}^{-1}). This treatment assumes proper weighting. If weighting was not used and unit weights were assumed, then the σ_{aj} are approximated by

$$\sigma_{aj} = s[B_{jj}^{-1}(w_i = 1)]^{1/2} \tag{5-40a}$$

$$s^2 = [1/(N - P - 1)] \sum [y_i - F(x_i)]^2 \tag{5-40b}$$

where s^2 is just the variance of the fit and B_{jj}^{-1} ($w_i = 1$) is the jth diagonal element of $\mathbf{B}^{-1}$ evaluated for unit weights.

A word of caution is in order if Eqs. (5-39) and (5-40) are used with the Marquardt method. If λ approaches 0 when the solution has converged, then the last matrix inverted is $\mathbf{B}^{-1}$. Frequently, however, λ is still significant at convergence. Thus, the inverse matrix of the modified $\mathbf{B}$ of Eqs. (5-39) and (5-40) can yield severe errors in the σ's. This problem is readily circumvented by allowing convergence, generating a B with $\lambda = 0$, and calculating $\mathbf{B}^{-1}$ for use in Eqs. (5-39) and (5-40).

The nonlinear least squares programs in Appendix E use Eqs. (5-39) and (5-40). Equation (5-39) with proper weights works very well. Equation (5-40), on the other hand, should be considered quite approximate. For example, a simulated SPC double-exponential decay ($K_1 = 9000$, $K_2 = 1000$, $\tau_1 = 35$, and $\tau_2 = 7$) was fit by nonlinear least squares with a sum of two exponentials. Twenty simulations were performed using both weighted and unweighted fits. The results are shown in Table 5-7. The observed results are the standard deviations calculated directly from the 20 calculated sets of K's and σ's using the normal definition of a standard deviation for a set of results. The calculated values were either from Eq.

TABLE 5-7

Comparison of Estimated and Observed Least Squares Fit Uncertainties[a]

	Weighted		Unweighted	
	Observed	Estimated with Eq. (5-39)	Observed	Estimated with Eq. (5-40)
	σ	$\sigma \pm \sigma_\sigma$	σ	$\sigma \pm \sigma_\sigma$
K_1	111	90 ± 44	111	27 ± 13
τ_1	0.20	0.21 ± 0.07	0.27	0.08 ± 0.03
K_2	107	97 ± 35	124	26 ± 12
τ_2	1.67	1.15 ± 0.44	1.35	0.27 ± 0.12

[a] Data of the form $9000 \exp(-t/35) + 1000 \exp(-t/7)$ with Poisson noise added before reduction. Each entry is the result of 20 simulations using nonlinear least squares fitting over the $0 \leq t \leq 99$ range.

(5-39) or (5-40) depending on whether the data fits were weighted or not. Also shown are the standard deviations of the calculated standard deviations σ_σ.

It is clear that, for the weighted fits, Eq. (5-39) works quite well and even a single experiment can provide the parameters and reasonable uncertainty estimates. For the unweighted case, however, the calculated uncertainties are unrealistically optimistic. For data with widely varying σ's, this appears to be a general problem of Eq. (5-40). Where proper weights are unavailable the formulation of Eq. (5-40) can be used to estimate relative uncertainties, but it should be used cautiously for stating error limits. Readers interested in a rigorous but complex treatment for two exponential fits should see the paper by Hinde *et al.* (1977).

G. FAST ITERATIVE EXPONENTIAL EVALUATION

A frequently required computation in many lifetime determinations, especially in least squares fitting for calculating χ^2, is evaluation of the series

$$T_i = a_1 \exp(-t_i/a_2), \qquad t_i = (i - 1)\,\Delta t, \qquad i = 1, \ldots, n \quad (5\text{-}41)$$

The most direct approach is to evaluate it n times using Eq. (5-41). Exponentials are slow functions to evaluate even with a computer, and in a mini- or microcomputer environment this computational time is significant. The following iterative formula greatly accelerates evaluations of Eq. (5-41):

$$T_{i+1} = a_1 \exp[-(t_i + \Delta t)/a_2]$$

$$T_{i+1} = a_1 \exp(-t_i/a_2)\exp(-\Delta t/a_2)$$

$$T_{i+1} = T_i C, \qquad i = 2, \ldots, n \qquad\qquad (5\text{-}42)$$

$$C = \exp(-\Delta t/a_2), \qquad T_1 = a_1$$

Thus, each succeeding term can be calculated from the previous term and a constant C. In Eq. (5-41), even if $-1/a_2$ were evaluated first, n exponential evaluations and $2n$ multiplications would be required to evaluate n terms. Equation (5-42) requires one initial division and exponential evaluation and then n multiplications. Regrettably, this recursion trick fails with other important decay schemes such as Förster transfer.

To compare the speeds of Eqs. (5-41) and (5-42) we evaluated 1000 terms using $\Delta t = 0.25$, $a_1 = 10{,}000$, and $a_2 = 40$. Using programs written in eight-digit North Star BASIC (2-MHz 8080-based microcomputer), Eq. (5-41) required 110 sec, whereas Eq. (5-42) took only 17 sec. On a

Hewlett-Packard HP-85 (12 significant figures), the times were 42 versus 9 sec. The program EXPFIT gains much of its speed by use of Eq. (5-42).

The recursive method is less accurate than the direct one because of the accumulation of propagated errors, but this is rarely a problem. In the above cases, the last term of the recursive formula was accurate to $10^{-3}\%$ in North Star BASIC and better than $10^{-7}\%$ on the HP-85. These discrepancies are insignificant, even though more terms were calculated than normally. For software that does arithmetic to about seven digits (i.e., 32-bit floating point representation), the errors are somewhat larger, but for 200–300 terms are still unimportant.

H. TESTING AND APPLICATIONS OF NONLINEAR LEAST SQUARES

There are a number of standard trial functions for testing nonlinear minimization routines. Several functions, along with contour maps and convergence data for different algorithms, are given by Daniels (1978). One of the most widely used is the Rosenbrock function

$$E = 100(x_1^2 - x_2)^2 + (1 - x_1)^2 \tag{5-43}$$

The accurate location of the minimum at (1,1) of this deceptively simple function is remarkably difficult. It is a steep-walled highly curved canyon with a relatively flat bottom. The usual starting point is at $(-1.2, 1)$. The grid search takes hundreds of iterations, and the simplex method takes 40 trials to reduce E to below 10^{-6}, but x_1 and x_2 are accurate only to 0.1 and 0.2%, respectively.

A good four-parameter function is Powell's function (Daniels, 1978). Numerous other test functions are available, but we have found the above two and the fitting of a noise-free and a noisy decay curve given by a sum of two exponentials to be good test functions.

An important application of nonlinear least squares is the solution of nonlinear equations. If we wish to find the roots of the nonlinear equation

$$F(x_1, x_2, x_3, \ldots, x_p) = 0 \tag{5-44}$$

a standard nonlinear least squares fit program can be used. All we have to do is guarantee that the solution is a minimum. This is easily done by solving for the minimum of

$$[F(x_1, x_2, \ldots, x_p)]^2 \tag{5-45}$$

If our minimum is zero, the x_i are a root.

This approach is also applicable to systems of nonlinear equations. Consider two nonlinear equations:

$$F_1(x_1, x_2, \ldots, x_p) = 0 \qquad (5\text{-}46a)$$

$$F_2(x_1, x_2, \ldots, x_p) = 0 \qquad (5\text{-}46b)$$

The solution of Eq. (5-46) is obtained by minimizing

$$F_1^2 + F_2^2 \qquad (5\text{-}47)$$

If the minimum of Eq. (5-47) is zero, the x_i are roots of Eq. (5-46). This approach is clearly generalizable to any system of equations.

6

Convolution Integrals

A. INTRODUCTION

So far we have solved a variety of interesting kinetic systems using the traditional approach. We write down the system's differential equation and then solve it. For example, we have solved for the important general case of an arbitrary excitation and an exponential system response. This direct approach has limitations: (1) The differential equations may not always be simple to solve. (2) Each kinetic scheme generally gives a new set of differential equations to be solved, which is rarely approached with enthusiasm, especially by a nonmathematical person. (3) It is also possible to invent decay schemes where one cannot write down the system's differential equation either simply or at all. (4) For tabular rather than analytical data, it may be necessary to solve the differential equations numerically, which can be fraught with difficulties. (5) Finally, the differential equation approach as commonly taught lacks overall elegance, generality, and simplicity.

We shall introduce the convolution integral concept which unifies, simplifies, and clarifies the mathematics of decay time measurements. We first develop the concept of the convolution integral in a nonrigorous intuitive manner and then show how it solves a host of differential equations by reducing them to much more tractable integral equations. Finally, we will demonstrate the utility of the convolution integral by solving problems previously handled by less elegant and powerful methods. In Chapters 7, 8, and 10, we use the convolution integral to prove fundamental and useful theorems that would be all but intractable by the classical methods of solution of differential equations.

B. DEVELOPMENT OF THE CONVOLUTION INTEGRAL

We first define the impulse response of a system $i(t)$ as the response that would be observed if we excited it with a pulse of zero duration but of finite energy. A unit impulse is an impulse that produces one unit of response. For example, we might define a unit excitation impulse as one that produces 1 mole or 1 mole/liter of excited states. Generally, however, we are interested only in relative responses and we can use relative impulses rather than unit ones. Now, given any arbitrary excitation pulse $E(t)$ and an $i(t)$, what would be the observed sample decay $D(t)$? Figure 6-1 shows a representative excitation pulse $E(t)$ that might arise from a slow flash and a sample impulse response that is a sum of two exponentials. What will $D(t)$ look like?

We will show that any continuous function can be approximated by a string of properly weighted and spaced impulses or spikes of measurable area but negligible width. If $E(t)$ is approximated by such a string of impulses, then because we know the impulse response for the system under study, we can calculate the $D(t)$ at different times as a simple summation of the contributions from all previous impulses. We first show how $D(t)$ can be obtained if $E(t)$ is a series of impulses. We then show how to decompose any continuous $E(t)$ into a string of impulses to which we then apply our previous formulation for obtaining $D(t)$. The resulting solution is the convolution integral.

This idea can be developed more fully by examining Fig. 6-2a where three excitation impulse spikes of different area contents are shown. The

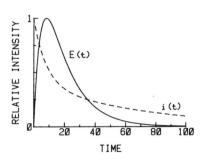

Fig. 6-1. Representative excitation pulse $E(t)$ and sample impulse response $i(t)$. The impulse response in this case is not a simple exponential but is given by $0.6 \exp(-t/8) + 0.4 \exp(-t/70)$.

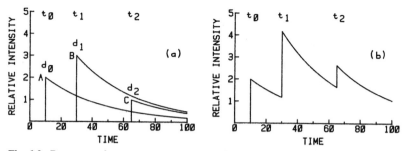

Fig. 6-2. Response of a sample with an exponential impulse response to a string of three impulses A, B, and C. (a) The three impulses are shown by the solid vertical lines at t_0, t_1, and t_2. The height of the lines (2, 3, and 1 at t_0, t_1, and t_2, respectively) represent the relative amplitudes of each impulse. The contribution of the sample decay from each impulse is shown by the three decay curves d_0, d_1, and d_2. (b) The observed response of the system to the string of impulses.

three impulses occur at t_0, t_1, and t_2 and have energy contents of 2, 3, and 1, respectively. We assume that $D(t) = 0$ for all $t < t_0$. We would like to calculate the response of this system which has a unit impulse response $\exp(-t/\tau)$. In this case one immediately sees that, for any time greater than t_2, the observed decay $D(t)$ will be the sum of the contributions from the three impulses at t_0, t_1, and t_2 [Eq. (6-1a)]. For times between t_1 and t_2, only the impulses at t_0 and t_1 contribute [Eq. (6-1b)] to the decay, and for times between t_0 and t_1 only the impulse at t_0 contributes. For times less than t_0 there is no decay, since the sample is not excited.

$$D(t) = d_0 + d_1 + d_2, \qquad t_2 \leq t \tag{6-1a}$$

$$D(t) = d_0 + d_1, \qquad t_1 < t < t_2 \tag{6-1b}$$

$$D(t) = d_0, \qquad t_0 < t < t_1 \tag{6-1c}$$

$$d_0 = 2 \exp[-(t - t_0)/\tau] \tag{6-1d}$$

$$d_1 = 3 \exp[-(t - t_1)/\tau] \tag{6-1e}$$

$$d_2 = 1 \exp[-(t - t_2)/\tau] \tag{6-1f}$$

where d_0, d_1, and d_2 are the responses of the sample to only the excitation pulses at t_0, t_1, and t_2, respectively. d_0, d_1, and d_2 are shown in Fig. 6-2a, and the resultant $D(t)$ given by Eq. (6-1) is shown in Fig. 6-2b.

In general, if there are a number of impulses at times t_j $(t_{j+1} > t_j)$ and each excitation impulse has an energy content or weight of E_j at t_j, then $D(t)$ for the exponential impulse response is

$$D(t) = \sum_{i=0}^{p} E_j \exp[-(t - t_j)/\tau] \tag{6-2}$$

where p assumes the largest value such that $t_p \leq t$. Further, note that there was nothing unique in the choice of an exponential $i(t)$. Thus, for any $i(t)$ the general equation is given by

$$D(t) = \sum_{i=0}^{p} E_j i(t - t_j) \tag{6-3}$$

where E and p have the same meaning as before.

We now confront the problem of a continuous rather than a discrete $E(t)$. A continuous $E(t)$ (e.g., Fig. 6-1) is decomposed into a series of impulses as follows: First, divide $E(t)$ into points taken at even times separated by Δt. We replace each continuous section of $E(t)$ between $t_j - \Delta t/2$ and $t_j + \Delta t/2$ with an impulse at t_j that has the same area $E(t)$ has over this section. In other words, we concentrate all the driving force from this interval into an equivalent impulse at t_j. If Δt becomes small enough, then we can estimate the necessary area under these segments E_j by

$$E_j = E(t_j) \, \Delta t, \quad j = 0, \ldots, N \tag{6-4}$$

We can now represent any continuous function as a string of excitation impulses E_j given by Eq. (6-4), which yields $D(t)$ from Eq. (6-3):

$$D(t) = \sum_{j=0}^{p} E(t_j) i(t - t_j) \, \Delta t \tag{6-5}$$

Since Δt is constant we can also write

$$D(t) = \Delta t \sum_{j=0}^{p} E(t_j) i(t - t_j) \tag{6-6}$$

Further, if we restrict our evaluations of $D(t)$ to the same times t_j at which we have $E(t)$ data, we obtain

$$D(t_p) = \sum_{j=0}^{p} E(t_j) i(t_p - t_j) \, \Delta t \tag{6-7}$$

where again the Δt may be removed for convenience. For small enough Δt Eq. (6-7) approaches the integral

$$D(t) = \int_0^t E(x) i(t - x) \, dx \tag{6-8}$$

After a change in variables this integral becomes (Kaplan, 1962)

$$D(t) = \int_0^t i(x) E(t - x) \, dx \tag{6-9}$$

Integrals of this form are *convolution integrals*. The convolution of two functions A and B is denoted by $A * B$ and is defined by

$$A * B = \int_0^t A(x)B(t - x)\, dx \tag{6-10}$$

The convolution integral finds wide application in the solution of differential equations such as arise in the area of linear electrical network analyses, engineering, and physics (Hauser, 1965; Kaplan, 1962), and many of these problems are equivalent to systems arising in luminescence work. Convolution integrals are both commutative and associative functions.

$$A * B = B * A \tag{6-11}$$

$$(A * B) * C = A * (B * C) \tag{6-12}$$

Using the definition of convolution integrals, we may rewrite Eqs. (6-8) and (6-9) as follows:

$$D(t) = E * i = i * E \tag{6-13}$$

Equation (6-9) follows directly from Eq. (6-8).

C. TREATMENT OF SPECIFIC SYSTEMS BY CONVOLUTION

A number of systems treated earlier in this book by classical methods of solving differential equations can now be handled extremely efficiently by use of the convolution integral. We demonstrate with two specific examples. In the general case, we begin by writing down $i(t)$, substitute it into Eq. (6-13), and carry out the necessary integrations.

We first solve the important case where we have an arbitrary excitation source $E(t)$ and an exponential impulse response $i(t) = K \exp(-t/\tau)$. Substitution of this $i(t)$ into Eq. (6-8) yields

$$D(t) = \int_0^t E(x)K \exp[-(t - x)/\tau]\, dx$$

$$= K \exp(-t/\tau) \int_0^t E(x) \exp(x/\tau)\, dx \tag{6-14}$$

This is, of course, the same as Eq. (4-3). A comparison of this one-line solution with the integrating factor approach of Appendix A leaves no doubt as to the efficiency of the convolution integral. Note that we did not even have to be able to write down the differential equation; all we needed to know was the system's impulse response. $i(t)$ was, of course, obtained by solving a differential equation, but it was a simple one.

It should also be noted that substitution into Eq. (6-9) is equivalent to Eq. (6-8); however, the resultant equation does not readily yield Eq. (6-14). Thus, care must be taken in deciding whether Eq. (6-8) or Eq. (6-9) is used in a specific problem.

Next consider the serial decay $A \to B \to C$ of Eqs. (4-1). Here [A] is written down directly, since it is just a simple exponential decay.

$$[A] = [A]_0 \exp(-k_{AB}t) \qquad (6\text{-}15)$$

To solve for [B], we recognize that the impulse response for B is

$$i_B(t) = \exp(-k_{BC}t) \qquad (6\text{-}16)$$

and that the pump or excitation for B is just [A]. Thus,

$$[B] = A * i_B = \int_0^t [A]_x \exp[-k_{BC}(x - t)] \, dx$$

$$= [A]_0 \exp(-k_{BC}t) \int_0^t \exp[(k_{BC} - k_{AB})x] \, dx$$

$$= [A]_0 \exp(-k_{BC}t)/(k_{BC} - k_{AB})[\exp[(k_{BC} - k_{AB})t] - 1] \qquad (6\text{-}17)$$

where $[A]_x$ is [A] evaluated at time x. Equation (6-17) rearranges to Eq. (4-7).

There are several important features of both of these examples. First, it was not necessary to write down elaborate differential equations. Second, it was not necessary to solve equations by traditional methods. As long as one can write down a model impulse response of the components, then one can proceed directly to the final stages of the solution, which is generally easier than the traditional method. Finally, one has reduced the solution of differential equations directly to the solution of integrals that are inherently easier to solve and more stable. For example, solutions via the convolution integral are straightforward and stable even if the functions are tabular arrays. Direct numerical solutions of differential equations, however, can be tricky, and instabilities can arise without warning. Additional examples of the power of convolution integrals are given in Chapters 7, 8, and 10. Should we encounter an intractable integral, however, recourse can always be made to numerical integrations which are simple to program and fast to carry out.

D. FAST ITERATIVE CONVOLUTION FORMULA

The defining relationship of the convolution integral can be used directly for calculating $D(t)$ from $i(t)$ and $E(t)$. Generally $E(t)$ is not an analytic function but is tabular data. In this case, the integrand of Eqs. (6-8) or (6-9) must be evaluated at every point and the integrations carried out numerically; the summations of Eq. (6-7) are commonly used. Direct application of these equations, however, requires a complete set of evaluations of the terms of the sum or integrand for every t. These computations can quickly become too slow to be useful. We describe important simplifications including the almost universally used fast iterative convolution formula. These simplifications are suitable only for $i(t)$'s that are exponentials or sums of exponentials, but this condition is by far the most commonly encountered.

If $i(t) = K \exp(-t/\tau)$, then Eq. (6-10) simplifies to Eq. (6-14). Now for each new time, only the additional contribution of the integrand from the last time to the current time has to be evaluated. All values of the integrand for earlier times, once evaluated, remain fixed. This yields an enormous increase in computational speed over direct application of Eq. (6-10). If $i(t)$ is a sum of n exponentials, then $D(t)$ is a sum of n terms each of the form of Eq. (6-14). We develop a recursive method of evaluating $D(t)$ as follows: Adopting the shorthand notation $D_j = D(t_j)$ we obtain

$$D_{j+1} = K \exp(-t_{j+1}/\tau)I_{j+1} \tag{6-18a}$$

$$I_{j+1} = I_j + \Delta I_j \tag{6-18b}$$

where

$$I_j = \int_0^{t_j} E(x) \exp(x/\tau) \, dx \tag{6-18c}$$

$$\Delta I_j = \int_{t_j}^{t_{j+1}} E(x) \exp(x/\tau) \, dx \tag{6-18d}$$

Thus, the integral for the $j + 1$ term can be calculated from the I_j and the additional term ΔI_j. This result is just a generalization of the fast exponential evaluation of Section 5.G.

Equation (6-18) can be simplified or enhanced still further by assuming a specific method for carrying out the integration of ΔI_j, particularly if the time interval between data points is constant. Demas and Crosby (1970) assumed that $E(t)$ across the interval t_j to t_{j+1} could be represented by a straight line connecting $E(t_j)$ and $E(t_{j+1})$; they then carried out the integrations exactly. This assumption yielded a recursive formula that reduced the number of data points required to give good calculated decay curves; they found it especially useful in treating hand-digitized oscilloscope photographs where a major goal was to reduce the number of points to be digitized and entered. Computationally, however, the Demas–Crosby equation is not particularly fast and is justified only when it is essential to minimize the number of data points.

By far the most popular recursive convolution formula was developed by Grinvald and Steinberg (1974). Their expression is now known as the fast iterative convolution or the Grinvald–Steinberg (GS) formula. They assumed that the interval between points was small enough so that trapezoidal rule integration across the interval was justified. In other words, they assumed that the area under a straight line connecting $E(t_j) \exp(t_j/\tau)$ and $E(t_{j+1}) \exp(t_{j+1}/\tau)$ gave a good estimate of ΔI_j in Eq. (6-18c). This approximation is similar to but not the same as the Demas–Crosby one, and many more data points may be required for accurate convolutions by the GS approximation. The trapezoidal rule yields

$$\Delta I_j = (\Delta t/2)[E_{j+1} \exp(t_{j+1}/\tau) + E_j \exp(t_j/\tau)] \qquad (6\text{-}19a)$$

$$\Delta t = t_{j+1} - t_j \qquad (6\text{-}19b)$$

Substituting Eq. (6-19) into Eq. (6-18) yields

$$\begin{aligned}
D_{j+1} &= K \exp(-t_j/\tau)I_j P + (K \Delta t/2)(E_{j+1} + E_j P) \\
&= D_j P + (K \Delta t/2)(E_{j+1} + E_j P) \\
&= [D_j + SE_j]P + SE_{j+1} \\
P &= \exp(-\Delta t/\tau), \qquad S = K \Delta t/2
\end{aligned} \qquad (6\text{-}20)$$

Equation (6-20) is exceptionally efficient for convolutions. Direct application of Eq. (6-18) is too slow to consider here. Even the use of Eq. (6-15) with a trapezoidal rule integration requires one new exponential evaluation and about six simple operations per calculated D_j. In the GS formula, however, after initial computation of P (involving one exponential) and S, only three multiplications and two additions are required for each D_j evaluated. This great economy of computational speed is essential for successful implementation of many deconvolution schemes such as deconvolution by iterative reconvolution (least squares).

The Grinvald–Steinberg formula has justly acquired wide acceptance. Sometimes overlooked, however, is the fact that it rests on a trapezoidal rule integration and is approximate. In particular, we noticed inaccuracies of Eq. (6-20) when we generated synthetic data for simulations. Examples of the errors that can arise are given in the following discussion. We used an $E(t)$ equal to a difference of exponentials with lifetimes of 5 and 6 nsec (Chapter 11). Table 6-1 shows a comparison of the $D(t)$'s computed exactly and by the GS approximation for τ's ranging from 0.2 to 20 nsec.

TABLE 6-1

Errors in the Fast Convolution Theorem (Percent Error in Calculated D(t)[a])

Time	E(t)	τ	0.2		2.0		5.5	10		20	
		Δt	1.0	0.5	1.0	0.5	1.0	1.0	0.5	1.0	0.5
0	0.0000										
1	0.0278		−203.3	−65.3	−10.83	−2.71	0.07	2.75	0.69	4.40	1.10
2	0.0462		−170.9	−53.6	−5.60	−1.40	0.02	1.37	0.34	2.17	0.54
3	0.0577		−161.6	−50.3	−3.87	−0.97	0.01	0.91	0.23	1.44	0.36
4	0.064		−157.1	−48.7	−3.02	−0.76	0.01	0.68	0.17	1.07	0.27
5	0.0667		−154.6	−47.8	−2.52	−0.63	0.01	0.54	0.13	0.85	0.21
6	0.0667		−152.9	−47.1	−2.19	−0.55	0.01	0.45	0.11	0.70	0.18
7	0.0648		−151.7	−46.7	−1.96	−0.49	0.00	0.38	0.09	0.60	0.15
8	0.0617		−150.8	−46.4	−1.79	−0.45	0.00	0.33	0.08	0.52	0.13
9	0.0578		−150.1	−46.2	−1.66	−0.42	0.00	0.29	0.07	0.46	0.12
10	0.0535		−149.6	−46.0	−1.56	−0.39	0.00	0.26	0.07	0.42	0.10
11	0.0491		−149.1	−45.8	−1.48	−0.37	0.00	0.24	0.06	0.38	0.09
12	0.0446		−148.8	−45.7	−1.42	−0.36	0.00	0.22	0.05	0.35	0.09
13	0.0403		−148.5	−45.6	−1.36	−0.34	0.00	0.20	0.05	0.32	0.08
14	0.0361		−148.2	−45.5	−1.32	−0.33	,0.00	0.19	0.05	0.30	0.08
15	0.0323		−148.0	−45.4	−1.28	−0.32	0.00	0.17	0.04	0.28	0.07
16	0.0287		−147.8	−45.3	−1.25	−0.31	0.00	0.16	0.04	0.27	0.07
17	0.0254		−147.6	−45.3	−1.22	−0.31	0.00	0.15	0.04	0.25	0.06
18	0.0225		−147.5	−45.2	−1.20	−0.30	0.00	0.14	0.04	0.24	0.06
19	0.0198		−147.3	−45.2	−1.18	−0.29	0.00	0.14	0.03	0.23	0.06
20	0.0174		−147.2	−45.1	−1.16	−0.29	0.00	0.13	0.03	0.22	0.06

[a] $E(t) = \exp(-t/5) - \exp(-t/6)$.

Representative values of Δt of 1.0 and 0.5 nsec were used; smaller Δt's with more points would, of course, yield smaller errors. Values of $E(t)$ are also given every nanosecond to permit visualization of the flash profile. The errors are especially severe for short lifetimes and $\Delta t = 1.0$ (>100%), although even for $\Delta t = 0.5$ the errors can be so large as to invalidate totally any convolutions by the GS formula. Even for long τ's and $\Delta t = 0.5$, the GS approximation can produce small but noticeable errors.

For the width of $E(t)$ (~13 nsec FWHM), Δt's of 0.5 and 1.0 are coarser than those that may arise with single-photon counting data. Consequently, in many cases the errors will be smaller than indicated. Our results stress the dangers of uncritical use of the fast iterative convolution formula. Thus, whenever the GS method is used in simulations or deconvolutions, especially with $\tau \leq \sim \Delta t$, it is essential to compare the resultant simulations with the exact calculations and deconvolutions of perfect noise-free data over the range of parameters to be encountered. We stress that these errors are an inherent characteristic of the GS method and are also present with noisy as well as noise-free data. Finally, the Demas–Crosby (1970) approach largely eliminates the errors, but at too great a price in computational speed to be attractive for single-photon-counting data where a high density of data is available.

CHAPTER

7

Real Detection Systems (and Does It Matter?)

A. REAL DETECTION SYSTEMS

So far we have treated the detection system as if it were ideal, and we have assumed that the observed $D(t)$ and $E(t)$ faithfully follow the true temporal dependence of the waveforms. This, of course, is not always possible. The detection–display system always distorts the waveforms. In some cases this distortion may be so low as to be unnoticeable, but in other cases the observed waveforms may be so badly distorted by the instrument that they carry virtually no information about the true waveforms.

Distortions are introduced by the detector, cabling, amplifiers, and display system. In this chapter we will consider a number of common sources of distortion. We will then describe how they affect mathematical deconvolution, and we will conclude with a discussion of the insidious effects of RC time constants.

We begin with distortions introduced by the detector. The most common detectors are photomultipliers (PMTs) and photodiodes. The interested reader is referred to the excellent phototube manual by RCA (1980) and the literature of other manufacturers such as EMI, Amperex, ITT, and Hamamatsu. Virtually all optical detectors are characterized by a time delay between the optical excitation and the appearance of the output signal. Further, even if the input light pulse is of zero duration (an impulse or delta function), the output pulse will have a finite pulse width. In a photomultiplier tube or vacuum photodiode the delay is caused by the finite time required for the ejected photoelectrons to traverse the internal space in the tube. In a PMT this involves impacting of the electrons on, and ejecting secondary electrons from, a series of internal dynodes. As not all photoelectrons, or secondary electrons, start out from the same place with the same velocity and direction, electrons follow different paths through the tubes and the signal pulses arrive at the collection anode at differing times and with differing distributions or spreads. For phototubes, these delays and spreads are known as the transit time and the transit time spread, respectively.

In a conventional multistage PMT the transit time can range from a few nanoseconds to ~100 nsec depending on a number of factors. Short path lengths and fewer stages of amplification reduce transit time. High accelerating potentials between stages increase acceleration on the electrons and reduce transit time. Thus, end-on PMTs with long path lengths between the photocathode and the first dynode tend to have long transit times. Compact side-view squirrel cage tubes tend to be very fast.

The very fastest, readily available tubes are planar vacuum photodiodes where the photocathode is a plane situated parallel, and as closely as possible, to a planar collection electrode. Very high accelerating potentials (typically 2–3 kV versus 100 V between stages for a normal PMT) are used. The high accelerating potential, the similar path lengths for all photoelectrons, and the short paths traversed make these tubes exceptionally quick with rise times of ~0.3 nsec.

The transit time, while it may be tens of nanoseconds, is not, however, a serious problem. It merely means that the electrical signals will appear delayed in time following the optical excitation pulse. But if all signals are delayed by the same amount [i.e., $D(t)$ and $E(t)$], then the only effect is a simple shift in the time origin of the experiment. Merely by redefining the origin one can then proceed as if there were no transit time delay. There are a few exceptions to this statement. In some cases the transit time depends on the wavelength of the optical pulse, in which case the shift between $E(t)$ and $D(t)$ differs and severe errors result. This complex and important topic is deferred to Section 10.B.

If different types of tubes are used in the same system, the large transit time shifts between the signals can produce an amusing (or puzzling to those encountering it for the first time) but not serious problem. For example, in our laboratory, we use an inexpensive robust RCA 931 side-view tube for monitoring our laser excitation pulse and triggering the detection electronics. The signal-monitoring tube is a much slower end-on IR-sensitive PMT. We find that the trigger pulse appears so much sooner than the sample signal that in the fastest sweeps of the oscilloscope the sample signal arrives after the sweep is complete. This problem is not serious, however, because the trigger pulse is easily delayed by inserting additional lengths of coaxial cable between the trigger tube and the oscilloscope. For example, the speed of the pulse in the cable is ~0.5–0.7 times the speed of light, which yields 1.5–2.0 nsec of delay per foot of cable added.

Far more damaging than transit time, however, is the transit time spread. This spread results in an overall degradation of pulse width and, thus, temporal resolution. The 10–90% rise times of observed output pulses for a delta function pulse excitation are one way of quantifying this spread. They can be as low as ~0.3 nsec for planar vacuum photodiodes and up to 5–20 nsec for slow end-on PMTs running at relatively low voltages. Thus, with a slow PMT, one might easily observe a 5-nsec-wide output pulse even if the input is of zero duration. We should also add that new microchannel plate (MCP) photomultipliers such as the ITT F4129 and F4143 have gains of 10^6 and rise times of <0.7 nsec; their current high prices, however, tend to restrict their use.

In SPC instrumentation the system impulse response is always shorter than that measured on an analog instrument. SPC instrumentation derives all its information from the rising edge of the pulses; the width is unimportant. The impulse response of the SPC instrument is limited by jitter in timing pulses and detection circuitry.

The distorted signal coming from the PMT is further distorted by the cable, wiring, and input characteristics of the monitoring system. These circuit elements all possess capacitance which must be charged by this current and inductance which can cause ringing. In a traditional measuring circuit (Fig. 7-1a), the phototube current is converted to a more easily measured voltage by connecting the collection electrode (anode) to ground through a load resistor. Unfortunately, as well as developing the signal, this load resistor also prevents the charged system capacitance from discharging instantly if there is an abrupt change in the signal. This storage stretches the observed voltage pulses beyond the actual current pulse. The characteristic discharge time is the RC time constant of this circuit. This storage characteristic of the capacitors also causes the voltage signal to lag behind the current for rising as well as falling currents. This distortion can be large enough to hide completely any information about the original optical pulse. We discuss this topic in greater detail in Section C.

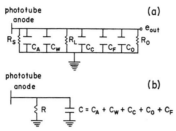

Fig. 7-1. Sources of resistance and capacitance in a phototube circuit (a) and the RC equivalent circuit (b). R_S, leakage resistance to ground; R_L, load resistance; R_O, input resistance of oscilloscope; C_A, capacitance of phototube anode; C_W, capacitance of phototube wiring; C_C, capacitance of cable connecting phototube to oscilloscope; C_F, capacitance added externally for increased filtering of the signal; C_O, input capacitance of oscilloscope; R, effective resistance ($1/R = 1/R_S + 1/R_L + 1R_0$); C, effective capacitance ($C = C_A + C_W + C_C + C_0 + C_F$).

Several important points are apparent in Fig. 7-1. For signals that vary slowly enough, the output voltage is proportional to R (= current $\times$ R). Thus, to maximize the signal it is advantageous to keep R as large as possible. Unless a preamplifier is used, R cannot exceed R_0, which is 1 M Ω for most regular oscilloscopes, but may be as low as 50 Ω for a fast detection system. As we will show, however, RC distortion frequently precludes taking full advantage of the maximum R.

Another point which is less obvious is that RC distortion can actually be used to advantage. The larger the RC time constant, the more slowly the signal can change across the load resistor. As long as one is not too greedy, this slowing of the response can be used to filter out high-frequency noise and enhance the signal-to-noise ratio. Thus, it is not uncommon to add capacitance to the measuring circuit (C_F) in order to slow down the response. We will return to this point later.

For very fast phenomena there can be signal reflections with concomitant ringing or slow rise times. Again we defer this discussion; see Section 10.E. for further details.

The display electronics add further distortions. The amplifiers cannot respond instantly to a step excitation and so exhibit finite rise times, and the rise time might be dependent on the gain. Further, the outputs of some amplifiers can change only at a maximum rate (i.e., a certain number of volts per microsecond); this characteristic is known as the slew rate limit of the amplifier. Thus, while such an amplifier can respond faithfully to small amplitude signals, it can distort large amplitude signals if they change more rapidly than the slew rate of the amplifier. Any amplifier can exhibit nonlinearities, especially at low or high amplitudes. Analog-to-digital converters can be nonlinear and exhibit only a finite step resolution determined by the number of bits. Oscilloscopes and oscilloscope cameras can exhibit barrel and pin cushion distortions. Oscilloscopes and transient digitizers can have nonlinear time bases, and TACs can be nonlinear.

With all these distortions it seems a wonder that valid data can be collected at all or that it can be reduced. Fortunately, most of these errors may be too small to be important, or calibration may be used to eliminate them.

The above-described errors can be divided into two classes: linear and nonlinear. A linear distortion is the same regardless of the input amplitude for all possible inputs. In other words, if the input is doubled, the output will double without any change in shape. PMT transit time and spread, RC smoothing, and high-frequency amplifier roll-off are all linear. Nonlinear distortions exhibit amplitude-dependent shapes. PMT saturation, slew rate limiting, and oscilloscope pin cushion distortions are all nonlinear.

Nonlinear distortions must be avoided or calibration must account for them. As we shall show, however, linear distortions can be eliminated in data treatment and without any specific knowledge of their form or magnitude.

B. LINEAR DISTORTIONS AND DECONVOLUTIONS

To date we have treated $D(t)$ and $E(t)$ as though they were true rather than observed quantities. We shall now show that, subject to one condition, all our previous results are correct even if we use observed, rather than true, quantities. The necessary condition is that all distortions must be linear and $E(t)$ and $D(t)$ must be measured under identical states of distortion. Therefore, it would not be acceptable to measure $E(t)$ at one PMT voltage setting and $D(t)$ at another setting, since the system responses are different.

Before beginning the derivation we define the following quantities:

$e(t)$ is the true time dependence of the excitation pulse,

$d(t)$ the true time dependence of the sample emission,

$i_s(t)$ the true time dependence of the sample emission when excited with a zero duration excitation pulse,

$i_d(t)$ the response of the optical electronic display system to a zero duration optical excitation pulse,

$E(t)$ the observed excitation pulse, and

$D(t)$ the observed emission decay.

Using the convolution integral (Chapter 6) we note immediately that the true sample decay is the convolution of the sample impulse response $i_s(t)$ with the true excitation pulse $e(t)$.

$$d(t) = e * i_s \tag{7-1}$$

The observed excitation and emission curves are the true excitation and emission curves convoluted with the impulse response $i_d(t)$ of the detection system:

$$E(t) = e * i_d \tag{7-2}$$

$$D(t) = d * i_d \tag{7-3}$$

By combining Eq. (7-1) with Eq. (7-3) and taking advantage of the commutative and associative properties of the convolution integral, we obtain

$$
\begin{aligned}
D(t) &= (e * i_s) * i_d \\
&= (i_s * e) * i_d \\
&= i_s * (e * i_d) = i_s * E \\
D(t) &= E * i_s
\end{aligned}
\tag{7-4}
$$

We see that Eq. (7-4) is identical to Eq. (7-1) except that observed rather than true quantities appear. Thus, any place where we previously used true quantities, we can now use observed quantities as long as we recognize that the distortions in both $E(t)$ and $D(t)$ must be linear and the same for both measurements.

C. *RC* TIME CONSTANT EFFECTS

Figure 7-1a is a schematic representation of a typical PMT detection system. The dominant sources of resistance and capacitance are shown. For operation up to perhaps 10 MHz, this circuit can be analyzed as the equivalent circuit shown in Fig. 7-1b. The equivalent resistance is just the parallel resistance for all current return paths to ground; generally the load resistor or the display system's input resistance is the most important (i.e., lowest) resistance. Similarly, the total charge storage ability is just the sum of all the parallel capacitances; cabling is usually a major contribution. This equivalent capacitor can store charge, and this charge can escape only through the equivalent load resistor R. For a coulomb packet of charge instantaneously deposited on the capacitor, this charge will decay exponentially with a time constant equal to RC, where RC has units of seconds if R is in ohms and C is in farads.

To put this problem into perspective, consider a PMT, cable, and oscilloscope system with $C = 150$ pF and $R = 1K$. This system is typical of normal wiring and cable capacitance and a load resistor used with moderate-strength signals. The RC time constant is 150 nsec. Thus, any phenomena occurring in much less than 150 nsec would be lost completely with the smearing-out effect of the RC network. It should be pointed out that many researchers have measured the RC time constant of their instruments and thought it was the true temporal response of the sample decay. Further, a smaller but not inconsequential number of workers have actually published RC time constants as valid sample lifetimes.

We will now consider in detail the effect of the RC time constant in systems where the sample decay would be exponential if it were not for the RC effect; that is, the flash has a much shorter duration than the sample lifetime. The voltage impulse response $i(t)$ of the cable and detector network to a unit charge excitation and the PMT output current from the sample $e(t)$ are given by

$$i(t) \propto \exp(-t/RC) \tag{7-5a}$$

$$e(t) \propto \exp(-t/\tau) \tag{7-5b}$$

where τ is the mean lifetime of the sample. If we now treat the sample emission $e(t)$ as an excitation pulse driving a sample with a response $i(t)$, we can use the convolution integral to solve for the apparent decay $D(t)$ of the composite system. Substitution of these responses into the convolution integral yields

$$D(t) = [I_0 R/(1 - Z)][\exp(-t/\tau) - Z \exp(-t/Z\tau)]$$
$$Z = RC/\tau, \quad Z \neq 1 \tag{7-6}$$

where I_0 is the maximum phototube current and $D(t)$ is the observed but RC-distorted sample decay. For convenience we have defined a dimensionless quality Z which is the ratio of the RC time constant to the sample lifetime. Unlike the original $e(t)$, these curves all rise relatively slowly to maxima before decaying.

The effect of RC distortion is demonstrated graphically in Fig. 7-2. A noise-free exponential decay with a lifetime of 200 nsec (a) is shown, as well as the effects of RC time constants ranging from 50 to 1000 nsec ($Z = 0.25$ to 5.0). The semilogarithmic plot of each curve versus time is also given. Figure 7-2 demonstrates several important points. All the RC-distorted curves rise to maxima before decaying, and the delay time to the maxima increases with Z. Even for a rather modest amount of RC effect of $Z = 0.25$, the RC distortion is substantial. The semilogarithmic plots

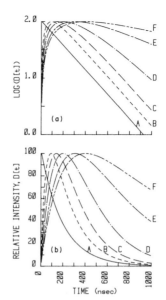

Fig. 7-2. Effect of *RC* distortion on an exponential decay curve with a mean lifetime of 200 nsec. (a) Linear plot of the data. (b) Semilogarithmic plot. A, undistorted $D(t)$; B–F, distorted $D(t)$'s with $RC = 50$ nsec (B), 100 nsec (C), 200 nsec (D), 500 nsec (E), and 1000 nsec (F).

are also revealing. For $Z = 0.25$ and 0.50, at long enough times, these plots linearize with slopes that are the same as the undistorted waveform, but the delay after the beginning of the decay is appreciable. Further, $D(t)$ has lost much of its peak amplitude before the semilogarithmic plots become linear. For example, with $RC = 50$ and 100 nsec, the semilogarithmic plots are visually linear after about 300 and 600 nsec, respectively; however, at these times the two decay curves are reduced to about 50 and 20% of their peak values. For the larger RC time constants ($Z > 1$), the semilogarithmic curves never approach the correct slope even at 1000 nsec. Indeed, for $Z = 2.5$ and 5, the curves are becoming linear, but with slopes corresponding to the 500- and 1000-nsec RC time constants of the measuring circuit.

We stress that the slow rise, peaking, and distorted decays are not caused by the sample lifetime but are instrument artifacts. Indeed, the functional form of Eq. (7-6) is identical to energy transfer or excimer decays, and some workers have erroneously misinterpreted such *RC*-distorted decays as arising from these more interesting effects.

The time at which the peaks occur t_{max} is readily determined by taking the derivative of Eq. (7-6), setting it equal to zero (for a maximum), and solving for the necessary time. The result is

$$t_{max} = \tau(Z \ln Z)/(Z - 1) \qquad (7-7)$$

where τ is the lifetime of the undistorted signal.

We take a brief side excursion here and demonstrate the beneficial effects of the *RC* time constant in data smoothing. Figure 7-3 shows how *RC* effects can both distort waveforms and enhance the signal-to-noise ratio of noisy input signals. The unfiltered signal (A) has a decay time of 20 μsec, and the signal-to-noise ratio is similar to what might be observed for a weak phosphorescence: Curves B–F in Fig. 7-3 show the effect of increasing *RC* time constants from 0.5 μsec to 8 μsec. Even for the very small *RC* = 0.5 μsec, which is 40 times shorter than the sample lifetime, a substantial reduction in the high-frequency noise greatly enhances the signal-to-noise ratio. C_F in Fig. 7-1 is filter capacitance added to raise *RC* in case the natural time constant is not large enough to give an adequate signal-to-noise ratio. As *RC* is increased, however, the rise time of the waveform increases just as in Fig. 7-2 for noise-free data. Even in the presence of considerable distortion, however, it is possible to measure good decay times (vide infra). We stress that the amount of information in the signal is actually reduced by filtering. If hand digitization of data is used and appropriate safeguards are employed for the amount of filtering, the increased ease of taking data from smoothed curves will more than compensate for the small loss of information.

We now grapple with the important questions: In the reduction of exponential decays from simple semilogarithmic plots of intensity versus time, how long an *RC* time constant can be tolerated before unacceptable errors result in the lifetime measurements? Is there some special way to take the data so as to minimize these errors? We first consider the case of a flash with a duration that is short compared to the decay time of interest.

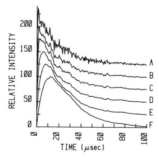

Fig. 7-3. Effect of *RC* distortion on a noisy exponential decay curve. Curve A is the noisy but undistorted curve with a mean lifetime of 20 μsec. Curves B–F are *RC*-smoothed (distorted) versions of curve A with *RC* = 0.5 μsec (B), 1.0 μsec (C), 2.0 μsec (D), 4.0 μsec (E), and 8.0 μsec (F). Each curve is displaced vertically from the others by 25 units to improve viewing. Smoothing was done digitally on waveform A rather than in an analog fashion, so that each curve could be compared directly with the original unsmoothed waveform.

Under these conditions we will obtain decay curves in the form of Fig. 7-2 or 7-3.

Inspection of Eq. (7-6) and Fig. 7-2 demonstrates that, as long as $RC < \tau$, it is possible to reduce or eliminate errors associated with the RC time constant. As t becomes larger, the exponential term associated with RC decreases faster than the term associated with τ. Thus, at long enough times, the decay curve will again become exponential with a characteristic lifetime equal to τ.

If $RC > \tau$, however, the longer the delay, the less the contribution the emission component will make. At long enough times, the decay will be exponential, but with a lifetime equal to RC. Under these conditions no direct extraction of τ from a semilogarithmic plot is possible.

We shall look at this problem by asking when the slope of the ln $D(t)$-versus-t curve equals the ideal $-t/\tau$ within some error bounds. $D(t)$ is given by Eq. (7-6), and the necessary derivative is given by

$$d \ln D(t)/dt = (-1/\tau)Y$$

$$Y = [1 - \exp(-Pt')/Z]/[1 - \exp(-Pt')], \qquad P = (1 - Z)/Z \qquad (7\text{-}8)$$

$$Z = RC/\tau, \qquad t' = t/\tau$$

Y is the term expressing the error in the correct slope of $-1/\tau$. If Y is unity, there is no error in the slope and, of course, as $Z \to 0$, $Y \to 1.000$. We have determined the necessary delay past the peak for Y to equal 0.9 (10% error), 0.95 (5% error), 0.99 (1% error), and 0.999 (0.1% error) by solving for t' in

$$Y - Q = 0$$

for $Q = 0.90, 0.95, 0.99,$ and 0.999.

Since it is clearly inappropriate to begin evaluating the slope before the peak, we determine how far past the peak the slope equals the specified error; t_{max} is given by Eq. (7.7). Table 7-1 summarizes the necessary delays past the peak (in units of τ) for the specified errors. Clearly, for small values of Z (e.g., 0.05), one need only delay the acquisition and reduction of data by 0.36τ to guarantee that the RC-induced error in the slope is less than 0.1%. Even if one were to start collecting data at 0.12τ past the peak, the initial points would yield a slope in error by no more than 10%. As Z increases, however, the necessary delay for a specified accuracy in the slope increases rapidly. For example, $Z = 0.5$ requires a delay of 1.7τ for a 10% error and a 6.6τ delay for a 0.1% error. Such delays may frequently be so large as to be unacceptable, since the signal amplitude is falling. However, Table 7-1 shows that, for modest Z's (<0.35), reasonably accurate τ's ($<10\%$ error) can be evaluated from semilogarith-

TABLE 7-1

Required Delay[a] (units of τ) for Given Errors[b] in RC-Distorted Decays

Z	Error (%) 10	5	1	0.1
0.05	0.12	0.16	0.24	0.36
0.10	0.25	0.32	0.50	0.76
0.15	0.38	0.50	0.78	1.19
0.20	0.53	0.70	1.10	1.67
0.25	0.68	0.91	1.44	2.21
0.30	0.85	1.14	1.82	2.81
0.35	1.04	1.40	2.25	3.49
0.40	1.24	1.68	2.73	4.27
0.45	1.46	1.99	3.29	5.16
0.50	1.70	2.35	3.92	6.22
0.55	1.98	2.76	4.67	7.47
0.60	2.29	3.23	5.56	8.99
0.65	2.64	3.78	6.64	10.88
0.70	3.05	4.44	7.99	13.31
0.75	3.54	5.25	9.75	16.57
0.80	4.12	6.27	12.14	21.21
0.85	4.84	7.64	15.66	28.43
0.90	5.78	9.58	21.50	41.53
0.95	7.06	12.69	33.88	>50.00

[a] Delay is past the peak.

[b] Errors in the slopes of the semilogarithmic plots versus time.

mic plots merely by delaying the acquisition about one lifetime τ past the peak of the decay curve. In reality, the errors would be even smaller than indicated, since Table 7-1 gives the slope error at a specific time, but data fitting would be done at still longer times where the errors are smaller. Finally, for very large Z's (>0.8), it is virtually impossible to evaluate accurate τ's from semilogarithmic fits.

We now turn to the question of how to eliminate RC errors in exponentials if they are too large to use the delay approach discussed above. The simplest approach is merely to reduce R or C until RC becomes acceptably small compared to τ. This is not always feasible. Signal strength is roughly directly proportional to R, and it may not be possible to reduce R without completely losing the signal.

Reducing C may be equally difficult. Coaxial cable has 25 pF of capacitance per foot, and it is rarely possible to reduce the cable length below 1 ft. The phototube and housing capacitance is typically 5 pF, and the transient digitizer or oscilloscope input capacitance is 25 pF. A probe with ~5–10 pF of capacitance may be used to eliminate the cable and oscilloscope capacitance, but generally a factor of 10 in sensitivity is lost. A preamplifier built into the PMT housing can greatly reduce the *RC* time constant, since the only capacitance across *R* is that of the PMT and preamplifier input. A current-to-voltage converter can be used even with long interconnecting cables, since the input resistance is much lower than the load resistor itself. The added cost, complexity, or loss of speed associated with the use of an amplifier may, however, preclude its use.

Thus, in many cases it is not possible to reduce *RC* experimentally or to delay the measurements to long enough times to eliminate *RC* errors. Under these conditions recourse must be made to mathematical reduction techniques. *RC* distortion is linear, and any of the standard deconvolution techniques that use $E(t)$ and $D(t)$ are fully acceptable as long as the *RC* time constant effect is the same for both measurements. Thus, nonlinear least squares, curve fitting, moments methods, and any of the transform methods would work.

For the case where the only appreciable distortion is from the *RC* time constant (i.e., $RC \gg$ excitation pulse width) simplifications are possible. Under these conditions the *RC* time constant can be measured with high accuracy merely by analyzing the observed exponential relaxation of $E(t)$ by the normal semilogarithmic plots of intensity versus time. Moments methods (MM) 1 and 2 (see Chapter 8 for details) require the center of gravity of the flash and the radius of gyration of the flash, but for a simple exponential decay the center of gravity of the $E(t)$ curve is RC and the radius of gyration is RC. Thus, using either moments method 1 or 2, respectively, we can evaluate τ as follows:

$$\tau = (\mu_1/\mu_0) - RC \tag{7-9}$$

$$\tau = [(\mu_2/\mu_0) - (\mu_1/\mu_0)^2 - (RC)^2]^{1/2} \tag{7-10}$$

where the μ_i are the *i*th moments of $D(t)$. Because of the difficulties in evaluating the higher moments accurately, Eq. (7-9) is more accurate than Eq. (7-10) if there is no time shift between the measurement of RC $[E(t)]$ and $D(t)$. If a time shift exists between the measurement of $E(t)$ and $D(t)$, as when the internal triggering mode is used on the transient recorder or oscilloscope, Eq. (7-10) eliminates the time shift at the expense of some loss of overall accuracy.

RC time constant effects can also distort exponential lifetime measurements made using a rectangular excitation pulse. Such a pulse is common for long phosphorescence decay measurements. A mechanical shutter is opened to illuminate the sample and closed after the emission intensity signal levels off. The decaying edge of this waveform is affected by RC in very much the same way as for the delta excitation discussed previously. Demas and Crosby (1970) have discussed in detail errors from this source. It turns out that, when one references all measurements to excitation termination, then the errors are comparable to those for the delta function excitation where measurements are made relative to the peak in $D(t)$.

The effect of RC time constants on a general flash when the flash width and RC are comparable has also been discussed. Demas and Crosby (1971) stated what they believed to be a conservative rule: "If the signal from the flash decays exponentially with a lifetime less than one half the observed decay time of the sample, then $RC/\tau = 0.1$ introduces no more than a positive 10% systematic error." The data must, however, have a good signal-to-noise ratio and be reduced after noticeable curvature is absent in the semilogarithmic plot versus time. In the general case where it is unclear whether a semilogarithmic plot is accurate or not, deconvolutions can always be performed to yield accurate lifetimes.

D. A SPECIAL WARNING ON THE MEASUREMENT OF RISE TIMES

A related RC problem arises in the measurement of excited state rise times. Rise time measurements have proved useful in the determination of fast relaxation phenomena. Consider the following system.

$$D + h\nu \rightarrow {}^{**}D \tag{7-11}$$

$$ {}^{**}D \overset{k_d}{\rightarrow} {}^{*}D \tag{7-12}$$

$$ {}^{*}D \overset{k_1}{\rightarrow} D \tag{7-13}$$

where $^{**}D$ represents an upper excited state species, a more stable conformer, or even a different chemical species than the relaxed but excited species $^{*}D$. A typical system might be where $^{**}D$ is an upper excited state and $^{*}D$ is the emitting level. If a short excitation pulse is used and the concentration of $^{*}D$ can be monitored by such a technique as its luminescence, then from the kinetics of the build up of $^{*}D$ one can determine k_d. In the common case where $k_d \gg k_1$, in the short time domain $[^{*}D]$ would rise exponentially to a pseudoconstant value (see Section 3.D). Then, by merely linearizing the appropriate plot, k_d could be determined directly. With luminescence to monitor $[^{*}D]$, this approach has been used to attempt to evaluate k_d, and in the case of rare earth emissions it has been successful.

The rise time method has one major pitfall in that it assumes no instrumental distortion. But the RC time constant of the measuring circuit can produce a rise time indistinguishable in appearance from a true rise time. Consider, for example, a system where $k_d = 0$ and the data are examined in a time interval short compared to $1/k_1$. Under these conditions a delta function flash creates a step function excitation profile in *D which in turn begins to emit at a constant rate. As shown in Section 3.D, however, a step emission signal falling on a detector whose response is governed by a single characteristic lifetime time constant (RC in this case) creates a detector response that rises exponentially to a steady state. The lifetime of the exponential growth is RC in this instance. Thus, an apparently good data set can give rise times that are only instrument responses and not sample properties. The number of competent workers who have overlooked this small fact is substantial.

Figure 7-4 shows a simulated rise time measurement which demonstrates the detrimental effect of too large an RC time constant. The sample τ was 20 μsec and $RC = 0.25$ μsec, which are typical of values that could arise in measurements of room temperature phosphorescence or of rare earth luminescence. The undistorted waveform (A in Fig. 7-4) exhibits noise but no detectable rise time; the slight falloff of signal is due to the relatively short time interval compared to τ. The RC distorted waveform shows an exponential rise and flattening characteristic of an actual slow excited state relaxation step, and if this were reduced by standard techniques, one would arrive at a rise time of $\tau \sim 0.25$ μsec. In the absence of an understanding of the measurement system, one would then quite incorrectly conclude that the fundamental excited state processes included a slow relaxation step. Figure 7-4 also demonstrates dramatically

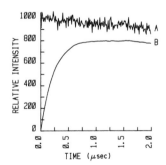

Fig. 7-4. Simulated effect of RC distortion on the rise time of a long-lived emission. Curve A is a noisy decay curve for an exponential decay with $\tau = 20$ μsec. Negligible instrumental distortion has been assumed. Curve B is curve A after passage through a system with an RC time constant of 0.25 μsec.

the very beneficial effects of a small amount of RC filtering ($Z = 0.0125$) in signal-to-noise enhancement.

The seriousness of the problem can be appreciated from the following examples. For fast decays a 50-Ω load resistor is used and C is rarely less than 50 pF. This combination yields $RC = 2.5$ nsec. Thus, using direct rise time measurements one could not see rise times less than about 10 nsec. For long-lived samples, load resistors of 10K are frequently required to obtain adequate signal. This combination yields rise times of 500 nsec. This value is not an unrealistic value for certain types of relaxation phenomena and, much to the chagrin of some workers, has been erroneously so interpreted in quality journals.

Thus, rise time measurements in particular should always be viewed with skepticism whenever finite values are found. The safest way to avoid this problem is to measure the RC time constant just by directly measuring $E(t)$ under the same conditions used to measure $D(t)$. If the excitation pulse is very short compared to RC, the output signal will then relax exponentially with the system RC time constant. If the observed RC time constant is comparable to the observed rise time, then the measurement is in error. Of course, if the observed excitation pulse width is comparable to the rise times of interest, regardless of its origin (e.g., a finite pulse width), the rise time measurements should not be attempted by the direct approach. It is also possible to calculate RC from estimates of R and C, but this is unwise when RC can be measured directly so easily.

The above warnings do not apply when deconvolution methods are used. Large RC time constants, however, degrade the limiting resolution because they stretch out the relevent waveforms.

CHAPTER

8

Deconvolution Methods

A. INTRODUCTION

In Chapter 5 we considered in detail the fitting of complex decays by means of least squares modeling. We ignored, however, the problem of extracting kinetic information when the flash duration is comparable to the sample's impulse response. We now address this problem.

Ideally, given $E(t)$ and $D(t)$, we would like to derive directly the sample impulse response $i(t)$. The process of extracting this $i(t)$ is called *deconvolution*. Once obtained, $i(t)$ can be fit to any kinetic model using least squares methods; however, for proper data analysis statistical information must be properly propagated. In practice, however, very few deconvolution methods yield $i(t)$ directly. Most actually assume a functional form for $i(t)$ (e.g., a simple exponential or a sum of exponentials) and then apply a deconvolution method that gives the best-fit parameters for this model. This procedure is also known as deconvolution.

Far and away the most important and commonly encountered deconvolution problems occur when a worker knows or has strong reason to suspect that $i(t)$ is a simple exponential or a sum of two exponentials. Therefore, most deconvolution methods strive to obtain the best kinetic parameters in an assumed $i(t)$ that is a single exponential or a sum of two or three exponentials. The fitting of a single exponential is probably the most common problem and is treated in depth here. As pointed out in

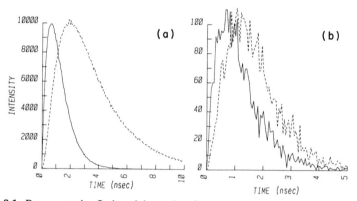

Fig. 8-1. Representative flash and decay data for deconvolutions. $E(t)$ ——. $D(t)$ ---. (a) has a lifetime of 2.5 nsec and exhibits typical single-photon counting noise, and (b) is for a 0.5-nsec lifetime and has noise characteristics of a poor analog instrument. [Reprinted with permission from Greer *et al.* (1981). Copyright 1981 American Chemical Society.]

Chapter 5, the fitting of a sum of two exponentials to a decay is tricky. The fitting of three exponentials is far more difficult and should be carried out only on the most carefully collected data. The additional deconvolution step further complicates this problem.

The range of experimental data researchers can encounter in deconvolution problems is illustrated in Fig. 8-1. In Fig. 8-1a the sample decay is long relative to the flash, and the signal-to-noise ratio is excellent; in this case it is typical of single-photon counting data. In Fig. 8-1b the lifetime is short [i.e., $\sim\frac{1}{3}$ the pulse width of $E(t)$] and very noisy; a very poor analog instrument or very weak sources might give these data.

We now discuss in detail a number of deconvolution methods. Unless explicitly stated we will deal with the fitting of a single-exponential impulse response. Limitations of space preclude covering interesting and valuable methods such as the exponential series method in which the impulse response is represented by a sum of exponential decays of fixed lifetimes and the preexponential factors are adjusted only to give the best least squares fit (Ware *et al.*, 1973; O'Connor *et al.*, 1979) and the method of modulating functions (Valeur and Moirez, 1973).

B. SIMPLE CURVE MATCHING

An exceedingly simple approach to the deconvolution of single exponentials is to take the measured $E(t)$ and compute a series of $D(t)$ curves for different assumed τ's. Then, by some form of curve matching, the observed decay is compared to these synthetic curves. The τ giving the closest visual match is assumed to be the correct lifetime. Gross visual discrepancies at the best fit are indications that the instrument is malfunctioning or that the sample impulse response is not a simple exponential.

The matching approach lends itself best to data where the height can be conveniently adjusted to match the height of the family of standard curves. Data stored in a transient recorder which can then be scaled at display time is one example. Another case is when a very high-repetition-rate stable source is used with a sampling technique. Under these conditions, trial runs can be quickly made to adjust the height. Lytle (1973), for example, used a sampling oscilloscope-based boxcar integrator and transparent overlays with the standard curves.

The matching approach is rapid and simple. It lacks high accuracy and the ability to model more complex kinetics. It does not seem to be widely used now that powerful inexpensive computer-interfaced deconvolution systems are available.

Although not strictly speaking a curve matching procedure, Rockley (1980) has described a nomogram method for rapid deconvolution of single exponential lifetimes. A calibration curve of the time interval between the $1/e$ time of the flash and the time where $D(t)$ decays to 0.075 of its peak value versus τ is generated synthetically. One then measures this time interval on the sample of interest and reads τ off the calibration curve. The approach worked remarkably well even for τ's half the width (FWHM) of $E(t)$. The calibration curve was relatively insensitive to the shape of $E(t)$ and would, thus, have to be generated only infrequently. This approach is, of course, less accurate than other approaches and gives no warning of complex decay kinetics.

C. ANALOG COMPUTATION

A somewhat more refined form of curve matching is the particularly ingenious method of single-exponential deconvolution by analog computation. The impulse response $i(t)$ of an RC network is $\exp(-t/RC)$. If an RC network is excited by a waveform $E(t)$, then the observed decay $D_{RC}(t)$ is obtained by substitution of $i(t)$ into Eq. (6-9):

$$D_{RC}(t) = K_{RC} \exp(-t/RC) \int_0^t \exp(x/RC)E(x)\, dx \qquad (8\text{-}1)$$

where K_{RC} is the proportionality constant. Comparison of Eqs. (4-3) and (8-1) reveals that, if $E(t)$ is fed into an RC network with a time constant RC, then the observed $D_{RC}(t)$ waveform exiting the network will be indistinguishable, except for scaling, from the true $D(t)$ if $RC = \tau$. Thus, all that needs to be done is to compare $D_{RC}(t)$ and $D(t)$ while varying RC until the two curves match visually. Then, if RC was dialed in with a calibrated RC network, τ can be read off directly.

For slow lamps and relatively long lifetimes, the approach described above would work. Complications arise in the nanosecond time domain, however, because of the difficulties in generating accurate RC time constants, cable ringing, and synchronization. In an ingenious fashion TRW Systems (TRW Systems, 1967; Mackey et al., 1965) solved these problems in a now obsolete and discontinued commercially available instrument. This system employed a dual-time base oscilloscope which could be used to display simultaneously the fast $E(t)$ or $D(t)$ and, in a much slower time domain, the simulated quantities. Because of the two time ranges, they simulated $E(t)$ by a low-speed dual RC time constant network driven by a pulse. The output shape of the simulated $E(t)$ was adjusted by varying the two time constants until the true and simulated $E(t)$'s matched as closely as possible on the oscilloscope. The $E(t)$ generator was, in effect, a difference of two exponentials and, in most cases, could be made to match the true $E(t)$ essentially perfectly as judged visually. This simulated low-speed $E(t)$ was then fed into the variable RC network, and the output was compared on the dual-trace oscilloscope with the true $D(t)$. The RC was adjusted to give the match with $D(t)$, and τ was read off. Even though the simulation was done in a much slower time domain, this represented merely a scaling factor which was readily accounted for. The principal errors were in the inability to exactly simulate $E(t)$ and in the visual matching process.

A major advantage of the TRW Systems simulator was ease and speed of operation. When a high-repetition-rate flash was used, the lifetime could be read off directly in a matter of seconds after inserting the sample. When the highest accuracy was not required and only single-exponential impulse responses were being examined, the system was excellent.

The development of computers and high-speed digital data reduction led to the demise of the analog deconvolution approach. It seems to the author, however, that transient digitizers and small, limited-function microcomputers may still benefit from this approach. Several very high-speed dual-memory transient digitizers are available. These permit two transients to be stored separately and displayed together on an oscilloscope. The speed of this display is fast enough to provide flicker-free waveforms but slow enough to permit analog computations. When $E(t)$ is stored in one channel and $D(t)$ in the other, all the elements necessary for successful analog simulation are available. One merely routes the recorded $E(t)$ through the RC network and compares it with $D(t)$; a small amount of digital logic and analog circuitry can do this easily.

In a related area, a sampling oscilloscope has been interfaced to a low-cost microcomputer to generate a digital boxcar integrator (Taylor *et al.*, 1980). This system permits the storing and displaying of two separate transients. Again, by recording $E(t)$ and $D(t)$ separately and then displaying and comparing the RC-distorted $E(t)$ with $D(t)$, analog simulation is possible. A major advantage of either of these systems over the TRW one is that, even though the simulation is done at low speeds, the experimentally observed $E(t)$ rather than an approximate function is being used.

D. PHASE PLANE METHOD

Demas and Adamson (1971) reported a transform method suitable for systems with single-exponential impulse responses. This technique linearizes the decay data, with the resulting plot having a slope equal to $-\tau$. Their original derivation was somewhat tedious, and a simpler version (Greer *et al.*, 1981) is given here. For a sample impulse response $i(t)$, given by $i(t) = K \exp(-t/\tau)$, the sample decay $D(t)$ is given by Eq. (4-3). Taking the derivative with respect to time yields

$$dD(t)/dt = K\left[E(t) + (-1/\tau) \exp(-t/\tau) \int_0^t E(x) \exp(x/\tau)\, dx\right]$$

$$= KE(t) - (1/\tau)D(t) \tag{8-2}$$

Integration of both sides between the limits 0 and t yields

$$D(t) = K \int_0^t E(x)\, dx - (1/\tau) \int_0^t D(x)\, dx \tag{8-3}$$

Rearranging yields two usable expressions:

$$Z(t) = -\tau W(t) + K\tau \tag{8-4a}$$

$$W(t) = (-1/\tau)Z(t) + K \tag{8-4b}$$

where

$$Z(t) = \int_0^t D(x)\, dx \bigg/ \int_0^t E(X)\, dx \tag{8-5a}$$

$$W(t) = D(t) \bigg/ \int_0^t E(x)\, dx \tag{8-5b}$$

Equation (8-4a) was the form originally proposed by Demas and Adamson (1981). For reasons discussed below, Eq. (8-4b) developed by Reed and Demas (1983) is the preferred form. In the first case a plot of $Z(t)$ versus $W(t)$ should be linear with a slope equal to $-\tau$ and an intercept equal to

$K\tau$. For Eq. (8-5b) a plot of $W(t)$ versus $Z(t)$ is linear with a slope of $-1/\tau$ and an intercept of K. In both cases the plots are known as phase plane (PP) plots. The necessary integrals are easy to evaluate by the trapezoidal rule.

For data taken at even time intervals Δt

$$\int_0^{t_j} D(x)\, dy = (\Delta t/2)R_j \tag{8-6a}$$

$$R_j = D(t_j) + D(t_{j-1}) + R_{j-1}, \qquad j \geq 1, R_0 = 0 \tag{8-6b}$$

and

$$\int_0^{t_j} E(x)\, dx = (\Delta t/2)Q_j \tag{8-7a}$$

$$Q_j = E(t_j) + E(t_{j-1}) + Q_{j-1}, \qquad j \geq 1, Q_0 = 0 \tag{8-7b}$$

These approximations yield

$$W(t_j) = 2D(t_j)/(\Delta t Q_j) \tag{8-8a}$$

$$Z(t_j) = R_j/Q_j \tag{8-8b}$$

The phase plane method has a number of advantages. The calculations are simple and trivial to program, and the computations are exceptionally quick. Both K and τ are obtained from the linear fit which then permits a calculation of $D(t)$ to verify the suitability of a single-exponential impulse response. Nonlinearity in the phase plane plot shows the kinetics to be more complex than for a simple exponential.

Figure 8-2 shows typical PP plots which, in this case, are derived from the data of Fig. 8-1. The first few points are inaccurate because of the large experimental errors in the $E(t)$ and especially $D(t)$. After the first few points, however, $W(t)$ and $Z(t)$ become well behaved. Even for the short-lived noisy data of Fig. 8-1b, the linearity of the PP plots is easily determined. The linearity demonstrates that $i(t)$ is characterized by a dominant single-exponential decay. The lifetimes calculated from these curves agree well with the true lifetimes (vide infra).

It seems at first glance that Eqs. (8-4a) and (8-4b) are equivalent. This is not true, however, when an unweighted linear least squares fit is used. Normal least squares fitting assumes that all errors are concentrated in the ordinate rather than the abscissa. Inspection of the form of $W(t)$ and $Z(t)$ reveals that $Z(t)$ is a ratio of integrals, whereas $W(t)$ is the ratio of $D(t)$ to an integral. Since integrals have signal-averaging properties, $Z(t)$ has smaller errors than $W(t)$, especially at later times. Therefore, especially for noisy data, Eq. (8-4b) is more nearly in accord with the assumptions of

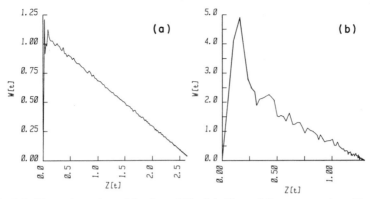

Fig. 8-2. Phase plane plots of the data of Fig. 8-1. Figure 8-2a corresponds to Fig. 8-1a, and Fig. 8-2b corresponds to Fig. 8-1b. [Reprinted with permission from Greer *et al.* (1981). Copyright 1981 American Chemical Society.]

unweighted least squares fitting. This is clearly shown in the PP plot of Fig. 8-2b where the dominant noise component is in $W(t)$ rather than $Z(t)$. Therefore, Eq. (8-4b) is definitely preferable to Eq. (8-4a). Reed and Demas (1983) first noted this problem and showed, using a number of digital simulations of single-photon counting data, that systematic errors in Eq. (8-4a) could be at least as high as 15–20%, but that Eq. (8-4b) was free of detectible errors. These authors also carried out extensive tests of the performance of the PP method with a wide range of lifetimes and noise levels in $E(t)$ and $D(t)$.

More recently Greer *et al.* (1981) carried out even more exhaustive tests. They used both Gaussian noise (noise level proportional to the square root of the channel counts) and true Poisson noise. They included a weighting factor in $W(t)$ to compensate for the increasing errors at shorter times. They also evaluated the sensitivity to fitting different regions (from the flash peak, from the decay peak, and the last 75% of their data). Additionally, they tested the suitability of the method for very short lifetimes.

We summarize the results of Greer *et al.* (1981): (1) There was no discernible difference between the use of Gaussian or Poisson noise. (2) The unweighted fit was equal to, or superior to, the weighted fit in accuracy and precision; the authors attributed this to the fact that $Z(t)$ also has errors that were not included in the simple least squares approach. (3) For reasonably noise-free data [≥ 1000 counts peak in $E(t)$ and $D(t)$], the accuracy and precision of the PP method is better than 1% for τ's longer than two-thirds of the FWHM of $E(t)$. (4) Using a 1.4-nsec-wide (FWHM) flash, they showed that, in principle, extremely short lifetimes (2.5 psec) could be extracted from data corresponding to accessible single-photon counting precisions [10^4–10^5 peak counts in $E(t)$ and $D(t)$]. (5) Visual inspection of even noisy data was good for evaluating the suitability of the single-exponential impulse model.

Jezequel *et al.* (1982) have generalized the phase plane method to the treatment of data that contain scattered excitation light or are characterized by an impulse response that is the sum of two exponentials. Love and Demas (1983a) have independently developed the scattered light correction. Derivation of the phase plane equation with scatter is given in Appendix D. Tests of the method using simulated decays show that, at least for the scatter correction, the PP method is fast and accurate even with very large scatter components.

Jezequel *et al.* (1982) pointed out a possible error source with the PP method when it is used with single-photon counting data. SPC data actually yield the integrated number of photons in a time interval rather than the instantaneous intensity. If the waveform changes too rapidly across the interval, the subsequent integrations of the PP method skew the PP curve and errors result. These errors are easily eliminated by using a high density of data points such as would normally be encountered in single-photon counting experiments. Jezequel *et al.* have shown mathematical approaches to the correction of such artifacts.

Love and Demas (1983b) have also applied the PP method to the extraction of decay parameters from Förster decay kinetics where the flash can be approximated as an impulse. Extensive simulations show the PP method to be fast and accurate. Further, they have shown that, theoretically, deconvolutions of decays exhibiting a Förster impulse response are also possible, although no simulations have been carried out yet. Bacon and Demas (1983) have extended the PP method to an exponential decay on an unknown baseline.

We summarize the status of the phase plane method. It is, with the possible exception of the simplified moments method, computationally the simplest and fastest deconvolution method. It is easily carried out on a hand calculator. Its accuracy is probably as good as that of any other single-exponential deconvolution approach. It gives good visual indications of whether or not the decay kinetics are simple. It is, thus, a very attractive method for routine deconvolutions. It can handle such complications as decay with stray light, baseline offset, or multiple-exponential decays. It has not been extended to account for a time shift between $E(t)$ and $D(t)$. In a suitably modified form it can be used to account for the wavelength dependence of the PMT transit time by means of a reference emitter (see Section 10.B.6). It also lacks proper use of statistical information inherent in the signal.

E. METHODS OF MOMENTS

1. General

Another popular deconvolution technique involves the moments methods which avoid some of the computational complexities of the nonlinear least squares. The moments methods take into account the shapes of $E(t)$ and $D(t)$ from their statistical moments:

$$\mu_k = \int_0^\infty t^k D(t)\, dt \qquad (8\text{-}9a)$$

$$m_k = \int_0^\infty t^k E(t)\, dt \qquad (8\text{-}9b)$$

where μ_k and m_k are the kth moments of the decay and flash, respectively.

The methods of moments were first developed by Bay (1950) for analyzing nuclear decay experiments. Cooper proposed their use in the deconvolution of fluorescence decay times, and Brody (1957) applied them to measurements of fluorescence lifetimes in his classic work on nanosecond flash fluorometry. Cooper (1966) subsequently analyzed the computational advantages of these approaches. More recently generalized formulations were developed by Isenberg and co-workers (Isenberg and Dyson, 1969; Isenberg *et al.*, 1973; Isenberg, 1973; Small and Isenberg, 1976, 1977, 1979, and 1983). Eisenfeld has published a number of papers on theoretical and practical implications of moments methods (Eisenfeld, 1979, 1983; Eisenfeld and Ford, 1979; Hallmark and Eisenfeld, 1979).

We first describe the two originally derived moments methods for evaluating a single decay time and then discuss the generalized method that permits evaluation of the K_i and τ_i for any decay curve that can be expressed as a simple sum of exponentials. A problem with Eq. (8-9) is that the integrals must be evaluated to infinite time, whereas in practice experimental data are taken to finite times. It then frequently becomes necessary to correct for the contributions beyond the experimental cutoff time. We discuss in detail several methods for eliminating these cutoff errors.

2. Moments Method 1 (MM1)

The simplest moments method for evaluating τ is to evaluate the center of gravity of $E(t)$ and $D(t)$ from

$$C_F = m_1/m_0 \qquad (8\text{-}10a)$$

$$C_D = \mu_1/\mu_0 \qquad (8\text{-}10b)$$

where C_F and C_D are the centers of gravity of $E(t)$ and $D(t)$, respectively. Then

$$\tau = C_D - C_E \qquad (8\text{-}11)$$

Remarkably τ is the difference between the centers of gravity of the decay and the flash. Qualitatively Eq. (8-11) is reasonable. The longer τ, the more $D(t)$ is shifted to longer times (Fig. 4-1) and the later the occurrence of the center of gravity.

Further, given the properties of the center of gravity, it is not even necessary to carry out these integrations. The centers of gravity can be found as is done in elementary physics laboratories for physical objects. Each curve is cut out and suspended by a string from an outside point. The suspending string points to the center of gravity. By repeating this process at two or more attachment points, the center of gravity can be obtained by triangulation. C_E and C_D can be read off and used in Eq. (8-11). (I am indebted to Dr. K. Wong for pointing out this clever method.)

3. Moments Method 2 (MM2)

The second approach to evaluating τ's from moments uses the radii of gyration of $E(t)$ and $D(t)$.

$$\sigma_E^2 = m_2/m_0 - (m_1/m_0)^2 \qquad (8\text{-}12a)$$

$$\sigma_D^2 = \mu_2/\mu_0 - (\mu_1/\mu_0)^2 \qquad (8.12b)$$

where the σ's are the radii of gyration. Physically, the radius of gyration comes out of elementary physics and corresponds to the distance from the center of gravity that all the object's mass could be placed so as to yield the same kinetic energy when the object is rotated. Then

$$\tau = (\sigma_D^2 - \sigma_E^2)^{1/2} \qquad (8\text{-}13)$$

Again, as with moments method 1, the result is qualitatively reasonable. The longer τ, the wider $D(t)$ and thus the larger its radius of gyration. Proofs of both the MM1 and MM2 equations presented by Brody (1957) are given by Demas and Crosby (1970).

Both Eqs. (8-11) and (8-13), in principle, yield good estimates of τ. Because of calculation errors in the higher moments, however, Eq. (8-13) is generally less accurate than Eq. (8-11). Further, the larger errors make MM2 less satisfactory for short lifetimes. For example, using oscilloscope data with a 20-μsec-wide (FWHM) flash, Demas and Crosby (1970) showed that 1-μsec lifetimes could be resolved with MM1 but only 2–3-

μsec lifetimes with MM2. MM2, however, has one important advantage over MM1: It is independent of any zero time shift between $E(t)$ and $D(t)$. The center of gravity depends on the position of the curve with respect to the origin. Thus, any time shift between the origins of $E(t)$ and $D(t)$ translates directly into τ errors. The radius of gyration, however, is independent of the origin, since it is defined around the center of gravity.

The advantages of the simple moments methods are that we do not have to make a subjective judgment of the goodness of fit of two curves, and the calculations are simple enough so that deconvolution can be done even on a mechanical calculator in <30 min (Cooper, 1966). The main disadvantage of these moments methods is that it is difficult to visualize the results, and unless the results of MM1 and MM2 are compared, there is no warning of a more complex $i(t)$ than the assumed single exponential. Alternatively, one can use the calculated decay parameters and calculate the expected decay curve, but the pristine simplicity of the method is lost. It is also sometimes difficult to obtain sufficiently accurate $D(t)$'s and $E(t)$'s at long times, especially if oscilloscope traces are used with MM2.

4. Generalized Moments Methods

If the sample impulse response is a sum of exponentials rather than a single exponential, the methods of moments can be generalized. For an impulse response given by

$$i(t) = \sum_{i=1}^{N} K_i \exp(-t/\tau_i) \tag{8-14}$$

we can define G_s by

$$G_s = \sum_{i=1}^{N} K_i \tau_i^s \tag{8-15}$$

It can then be shown (Isenberg and Dyson, 1969) that

$$\mu_0 = G_1 m_0, \qquad \mu_1 = G_1 m_1 + G_2 m_0$$

$$\mu_2/2! = G_1(m_2/2!) + G_2 m_1 + G_3 m_0$$

$$\vdots \tag{8-16}$$

$$\mu_k/k! = \sum_{s=1}^{k+1} ([m_{k+1-s}/(k + 1 - s)!]G_s$$

This is a linear set of equations in the G's. If there are N exponentials, then with $2N$ equations one can solve for $G_1, G_2, \ldots, G_{2N}$. It can then be shown that the desired $\tau_1, \tau_2, \ldots, \tau_N$ lifetimes are the roots of the polynomial equation

$$
\begin{vmatrix}
1 & \tau & \tau^2 & \ldots & \tau^N \\
G_1 & G_2 & G_3 & \ldots & G_{N+1} \\
G_2 & G_3 & G_4 & \ldots & G_{N+2} \\
& & \vdots & & \\
G_N & G_{N+1} & G_{N+1} & \ldots & G_{2N}
\end{vmatrix} = 0 \qquad (8\text{-}17)
$$

Solution of this polynomial equation yields the N values of τ. These τ's are substituted into the following equations:

$$
\begin{aligned}
G_1 &= \Sigma \, K_i \tau_i \\
G_2 &= \Sigma \, K_i \tau_i^2 \\
&\vdots \\
G_N &= \Sigma \, K_i \tau_i^N
\end{aligned} \qquad (8\text{-}18)
$$

where the summations run from 1 to N. Solution of these N linear equations yields the K_i in Eq. (8-14). For $N = 1$ moments method 1 is derived from Eqs. (8-16) and (8-17).

5. Moments Index Displacement (MD)

Less obvious than the basic moments methods is that, in fact, there are an infinite number of sets of equations that can yield the lifetimes. The necessary equation (Isenberg, 1973) is

$$
\begin{vmatrix}
1 & \tau & \tau^2 & \ldots & \tau^N \\
G_{1+n} & G_{2+n} & G_{3+n} & \ldots & G_{N+1+n} \\
G_{2+n} & G_{3+n} & G_{4+n} & \ldots & G_{N+2+n} \\
\vdots & \vdots & & \vdots & \\
G_{N+n} & G_{N+1+n} & G_{N+1} & \ldots & G_{2N+n}
\end{vmatrix} = 0 \qquad (8\text{-}19)
$$

where n is the moments displacement index and can assume any value from 0 to infinity. Comparison of Eqs. (8-17) and (8-19) shows that a moments displacement index (MDI) of zero ($n = 0$) yields the normal generalized moments equation. The highest moment that must be evaluated is $2N + n$. Thus, for a two-exponential fit and a MDI of 1, moments through the fifth must be evaluated.

Because of the rapidly escalating errors in evaluation of the higher moments, n should be kept small. Indeed, if it were not for the unique features of MD with $n \geq 1$, there would probably be no reason for using MD at all. For $n \geq 1$, there is no error associated with an arbitrary time shift between $E(t)$ and $D(t)$. In this regard MD is like moments method 2. Further, any MDI of 1 or greater shows greater insensitivity to flash variability than the standard moments method. Additionally, scattered light in $D(t)$ does not affect the data reduction. Numerous examples of the use of MD for the analysis of up to three component systems with time shift and stray light are given by Small and Isenberg (1976, 1977, 1983). It is clear that, in the hands of workers experienced in its use, MD is capable of quite remarkable feats of correcting for scattered light and time shift. For example, Small and Isenberg (1977) show beautiful examples of the extraction of three lifetimes (1.55, 7.84, and 18.88 nsec) in a mixture with both scatter and a 0.1-nsec time shift. The standard moments method yields errors of up to a factor of 4, and a MDI of 1 yields values typically good to 1–2%.

6. Cutoff Corrections

We now analyze the errors associated with the contribution to the moments of the data beyond the experimental cutoff time. This problem is called the cutoff correction. The methods for treating the cutoff correction are instructive, as they serve as models for other systems in which data are required beyond the region of quality data. We discuss here the correction for a single-exponential decay, but the extensions are straightforward with the equations given by Isenberg and Dyson (1969).

The m's and μ's given in terms of the cutoff time T are given by

$$\mu_k = {}^0\mu_k + \delta\mu_k, \qquad m_k = {}^0m_k + \delta m_k$$

$$^0\mu_k = \int_0^T t^k D(t)\, dt, \qquad {}^0m_k = \int_0^T t^k E(t)\, dt \qquad (8\text{-}20)$$

$$\delta\mu_k = \int_T^\infty t^k D(t)\, dt, \qquad \delta m_k = \int_T^\infty t^k D(t)\, dt$$

where the superscript 0 denotes the values of the moments evaluated up to the cutoff time and the $\delta\mu_k$ and δm_k are the contributions or corrections beyond T. The goal of any cutoff correction is to evaluate the $\delta\mu$'s and δm's.

We first show the extent of the problem. Figure 8-3 shows a typical

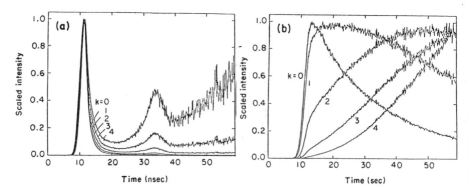

Fig. 8-3. The functions $t^k E(t)$ (a) and $t^k D(t)$ (b) for single-photon counting data without the use of exponential depression. [From Small and Isenberg (1983).]

single-photon counting $E(t)$ and $D(t)$ as well as the integrands of Eq. (8-9) for $k = 0$ to 4. With $k = 0$, the integrands are just $E(t)$ and $D(t)$. Especially for the higher moments with $D(t)$, it is clear that the cutoff time will produce severe errors. Indeed, for $k > 1$ with $D(t)$ the integrand has not even peaked, much less decayed, to the zero required for the integration over this region to be acceptable.

The simplest data handling method is just to be sure to take data far enough out on the trailing edge of $D(t)$ and $E(t)$ so that the $\delta\mu_k$ and δm_k are negligible. This approach can work very well for the zeroth and first moments required by MM1. It becomes increasingly less reliable for higher moments. It is, in general, a disaster for $k \geq 3$.

Probably the most obvious true correction method, and a reasonable approach for MM1 and MM2, is extrapolation. Demas and Crosby (1970) noted that both their $E(t)$ and $D(t)$ were approximately exponential at long times. They fit the tailing portion of each curve to an exponential using a standard linear least squares fit to the semilogarithmic plot of intensity versus time. They then evaluated the $\delta\mu$'s and δm's using these exponential decays and exact integration; see the original reference for details.

A more sophisticated approach is iteration. First we assume that $E(t)$ is zero for $t > T$, which means that all the δm_k will be zero. Further, $D(t)$ will decay exponentially with its characteristic lifetime for $t \geq T$. The $\delta\mu_k$ are then given by

$$\delta\mu_k = D_T \int_T^\infty t^k \exp(-t/\tau) \, dt \qquad (8\text{-}21)$$

where D_T is the effective amplitude of $D(t)$ at time T and τ is the sample lifetime. Equations (8-21) are exact integrals which are most conveniently calculated by recursion:

$$\delta\mu_0 = D_T\tau \exp(-T/\tau) \tag{8-22a}$$

$$\delta\mu_k = \tau[T^k \exp(-T/\tau) + k\,\delta\mu_{k-1}], \qquad k \geq 1 \tag{8-22b}$$

Thus, $\delta\mu_1$ can be calculated from $\delta\mu_0$; and then $\delta\mu_2$ can be calculated from $\delta\mu_1$, etc.

The iterative process proceeds as follows: One calculates τ using the uncorrected $^0\mu_k$ to yield an estimated $^0\tau$. Then $^0\tau$ is used in Eq. (8-22) to yield an estimate of the $\delta\mu_k$ denoted by $^0\delta\mu_k$. New μ_k are calculated:

$$^1\mu_k = {^0\mu_k} + {^0\delta\mu_k} \tag{8-23}$$

Carrying through the τ calculation using the new improved μ_k yields an improved guess of τ ($^1\tau$) which can be used to improve the correction in Eq. (8-22). This process of correcting the μ's is iterated using

$$^{j+1}\mu_k = {^0\mu_k} + {^j\delta\mu_k} \tag{8-24}$$

until the difference between $^j\tau$ and $^{j+1}\tau$ falls below some specified error threshold. This formidable looking procedure is in fact simple. The $^0\mu_k$ need be evaluated only once and the $^j\delta\mu_k$ can all be expressed in simple closed algebraic form.

Table 8-1 shows the iterative deconvolution of a single-exponential decay using the moments method 1 and moments method 2. Decay times of 5–50 nsec were used. A cutoff of 20 nsec was selected to produce large cutoff errors for the longer lifetimes. We have assumed the simplest type of cutoff scheme. We assume D_T is $D(T)$. For $\tau = 5$ nsec convergence by MM1 and MM2 was quite quick. Because MM1 used only first moments, whereas MM2 uses second moments, one would expect that MM2 would converge more slowly. This is clearly seen for $\tau = 25$ and 50 nsec where cutoff corrections are quite large.

Table 8-2 shows the results of deconvoluting 100 simulated decay curves by moments methods 1 and 2 for different peak counts in the $E(t)$ and $D(t)$ curves and for the lifetimes of Table 8-1. Data with peak counts of 10^4 will have noise levels like those of Fig. 8-1a, whereas those with peaks of 100 will resemble the data of Fig. 8-1b. The mean and standard deviations are given. It is clear that, except for the very noisiest data, the deconvolutions are quite acceptable. MM2, which uses the second moment, clearly gives a higher degree of uncertainty.

We have assumed that the cutoff correction for the flash can be disregarded, but is this so? Since T is the same for $E(t)$ and $D(t)$, it really does not matter whether the flash actually extends beyond T. When we ignore the cutoff correction for the flash, all our calculations for the decay curve are exact, as if the flash actually did terminate at T. That the flash continued is irrelevant. This truncation of $E(t)$ could, however, be carried to an

TABLE 8-1

Effect of Iterating in Lifetime Calculations by the Moments Methods 1 and 2[a]

τ Method j[b]	5		25		50	
	1	2	1	2	1	2
0	4.951	4.980	22.067	22.828	31.245	31.261
1	4.952	5.004	24.485	24.801	41.529	37.167
2			24.816	24.358	45.953	41.049
3			24.863	24.539	47.988	43.741
4			24.869	24.598	48.949	45.667
5			24.870	24.617	49.405	47.070
6				24.623	49.630	48.107
7				24.625	49.736	49.9880
8				24.626	49.788	49.459
9					49.813	49.896
10					49.825	50.226
11					49.830	50.477
12					49.833	50.667
13					49.834	50.812
14					49.835	50.922
20						51.207
25						51.259
30						51.273
33						51.276

[a] All results were calculated for simulated single-photon counting data with $E(t)$ given by Eq. (11-1) using $A = 5.5$ nsec and $B = 6.6$ nsec. Peak flash and decay channels were adjusted to 10,000 counts before adding noise.

[b] Iteration. Note discontinuities in iterations listed for moments method 2 and the $\tau = 50$ nsec data.

illogical extreme. One could throw away perfectly good flash and decay data by picking too short a T. This procedure would still be mathematically acceptable, but it would reduce the overall accuracy of the final result.

The above iterative procedure has, in fact, a potentially significant error built in. We have assumed that $D(t)$ is exact. In reality, this point (like all others) has noise on it. Thus, we are basing out cutoff correction on one potentially inaccurate point. For moments methods 1 and 2 used with reasonable quality data, the errors are usually negligible, since moments only through the second are required. The generalized moments method, however, uses higher moments and is very sensitive to cutoff correction errors (Fig. 8-3). It, thus, becomes essential to minimize these errors. This point could be found more reliably by fitting a straight line

TABLE 8-2

Accuracy and Precision of Moments Methods Lifetime Calculations[a]

τ	Method	N_{peak}[b]		
		100	1000	10,000
5	1	5.036 ± 0.152	4.982 ± 0.065	4.997 ± 0.019
	2	4.855 ± 0.477	4.992 ± 0.131	5.004 ± 0.037
25	1	26.012 ± 2.029	24.952 ± 0.471	25.020 ± 0.093
	2	25.576 ± 2.730	25.335 ± 1.195	24.974 ± 0.308
50	1	50.668 ± 6.489	50.529 ± 1.592	50.115 ± 0.644
	2	52.258 ± 8.810	47.959 ± 3.157	49.776 ± 1.075

[a] Except for the peak flash and decay counts, the conditions and methods are the same as for Table 8-1.

[b] The flash and decay peaks are equal before the addition of noise. Indicated errors are the mean and standard deviation of 100 simulations.

through the last few points and calculating the value on this line at T, which is similar to the approach used by Demas and Crosby (1970). A computationally more complex but better approach is available. Isenberg *et al.* (1973) calculated $D(T)$ from the entire preceding $E(t)$ curve. In the generalized moments methods the preexponential factor K is obtained for the impulse response and $D(T)$ can be calculated directly from Eq. (6-14) by numerical integration from 0 to T. This trick avoids putting excess weight on the observed noisy $D(T)$ and, in effect, signal-averages by computing what $D(t)$ would be by using the entire $E(t)$.

7. Exponential Depression

Most transform methods need to have data to infinite time; this is frequently not possible with the result that the accuracy and convergence properties vary. An ingenious solution to the problem is to depress the long-time data so that it has smaller effects on the transform, but in such a way that the desired decay parameters can still be extracted from the resultant data. If we multiply both sides of the convolution integral [Eq. (6-9)] by $\exp(-\lambda t)$, where λ is a positive depression parameter,

$$e^{-\lambda t}D(t) = \int_0^t [\exp(-\lambda t)]E(t-x)i(x)\,dx$$

$$= \int_0^t \{\exp[-\lambda(t-u)]\}E(t-u)[\exp(-\lambda u)]i(u)\,du \quad (8\text{-}25)$$

If we define

$$D_\lambda(t) = [\exp(-\lambda t)]D(t) \qquad (8\text{-}26a)$$

$$E_\lambda(t) = [\exp(-\lambda t)]E(t) \qquad (8\text{-}26b)$$

$$i_\lambda(t) = [\exp(-\lambda t)]i(t) \qquad (8\text{-}26c)$$

then

$$D_\lambda(t) = \int_0^t E_\lambda(t - x)i(x)\,dx \qquad (8\text{-}27)$$

Thus, the convolution integral is invariant with respect to exponential depression of the data. If $D_\lambda(t)$ and $E_\lambda(t)$ are deconvoluted by any method, one obtains not $i(t)$ but $i_\lambda(t)$ which can readily be corrected by Eq. (8-26c) back to the original lifetimes.

Figure 8-4 shows the effect of exponential depression on the data of Fig. 8-3. The enormously reduced cutoff corrections are evident. The reduced cutoff correction has the very beneficial effect of reducing the number of iterations required for convergence or forcing convergence of nonconverging systems. For example, Isenberg *et al.* (1973) present extensive results showing the frequently dramatic effects of exponential depression. In one case they attempted to fit a two-exponential decay with three exponentials using no depression. After 500 iterations there was no convergence. With exponential depression only six iterations were required.

In addition to speeding convergence, exponential depression also improves accuracy by suppressing noisy long-time information. It must be

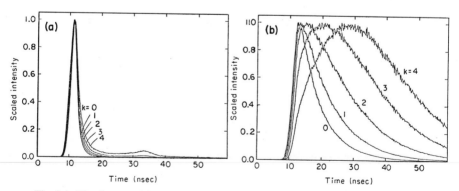

Fig. 8-4. The depressed functions $t^k E(t)$ (a) and $t^k D(t)$ (b) for the data of Fig. 8-3, where $\lambda = 0.10$ nsec^{-1}. The enormously beneficial effects of exponential depression especially for the higher moments of $D(t)$ are clearly evident. [From Small and Isenberg (1983).]

used with caution, however, as too much depression can result in a loss of information about long-lived components.

F. LAPLACE AND FOURIER TRANSFORM METHODS

Laplace and Fourier transforms are standard methods of solving differential equations and of analyzing electrical networks. It is not surprising, therefore, that they have found applications in fluorescence lifetime measurements. We describe here briefly the use of both these transformation approaches and their strengths and weaknesses.

We begin with the method of Laplace transforms. This approach was used by Helman (1971). It was then extended by Gafni et al. (1975). Our discussion follows the latter authors.

The Laplace transform $M(s)$ of a function $M(t)$ is given by

$$M(s) = L[M(t)] = \int_0^\infty M(t) \exp(-st)\, dt \qquad (s > 0) \qquad (8\text{-}28)$$

The Laplace transform is a linear operator

$$L[aM(t) + bN(t)] = aM(s) + bN(s) \qquad (8\text{-}29)$$

The transform of an exponential is

$$L[A \exp(-kt)] = A/(s + k) \qquad (8\text{-}30)$$

where k is the reciprocal of the sample lifetime. The Laplace transform of the convolution integral Eq. (6-10) is given by

$$D(s) = L[D(t)] = L[E * i] = E(s)i(s) \qquad (8\text{-}31)$$

where for the current discussion $E(t)$ and $D(t)$ are the observed excitation and the sample decay, respectively, and $i(t)$ is the sample impulse response. Thus, the complex process of convolution in the time domain converts to a simple product in the s domain. Equation (8-31) can be rearranged to

$$i(s) = D(s)/E(s) \qquad (8\text{-}32)$$

where simple division in the s domain gives the impulse response in the s domain. Ideally, one would like to take the inverse Laplace transform and convert $i(s)$ to the time domain, but such inverse transforms are not, in general, easily performed. Therefore, parameter fitting has been done in the s domain. For $i(t)$ equal to a sum of exponentials

$$i(s) = D(s)/E(s) = \Sigma\, K_i/(s + k_i) \qquad (8\text{-}33)$$

where the K_i and k_i are preexponential factors and decay rate constants for the ith component, respectively. The summation runs over the number of components. Thus, in principle, for the solution of a sum of N exponentials, one merely evaluates $D(s)/E(s)$ for $2N$ values of s and solves for the N values of K_i and the N values of k_i in the resultant $2N$ nonlinear equations. Alternatively, for a single exponential, a plot of $i(s)^{-1}$ versus s should be linear with a slope of k^{-1} and an intercept of k/K. Proper weighting is, however, not readily carried out.

As with the moments methods, a problem immediately arises. The necessary transforms are integrals to infinite time, whereas the data are to finite times. There is, thus, a cutoff correction. As with the moments methods an iterative solution is used to make the cutoff correction. Initially, one generates an approximate set of K_i and k_i using the truncated data. These K_i and k_i are then used to generate corrections yielding improved $i(s)$ which are then used to generate new K_i and k_i. The process is repeated until a self-consistent set of parameters results. Typically three to five iterations are required. Detailed equations are given by Gafni *et al.* (1975).

Gafni *et al.* (1975) tested this approach on real and simulated data. In particular, they present one very interesting result. They calculated the accuracy of deconvoluted parameters for a sum of two exponential decays as a function of $\tau_2/\tau_1 (K_1 = K_2)$. Figure 8-5 summarizes their results. For $\tau_2/\tau_1 > 2$ an accuracy of better than 1% was obtained, but for $\tau_2/\tau_1 < 1.6$ the errors in τ_2 increase extremely rapidly. Although obtained by the

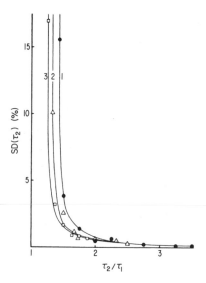

Fig. 8-5. The standard deviation obtained for τ_2, the longer of the two decay constants of synthetic double-component decay curves as a function of the ratio of the two decay constants. Curve 1, $\tau_1 = 4$ nsec; curve 2, $\tau_1 = 6$ nsec; curve 3, $\tau_1 = 8$ nsec. The flash width is about 10 nsec. [From Gafni *et al.* (1975).]

Laplace method, the results are similar to those observed in simple linear least squares fitting and least squares deconvolution.

The Laplace transform method can also be generalized to handle stray light and PMT time shift. For example, if there is stray light in the signal, $D(t)$ becomes

$$D(t) = CE(t) + E * I \qquad (8\text{-}34)$$

where C is the scatter coefficient. Taking the Laplace transform yields

$$D(s) = CE(s) + E(s)I(s) \qquad (8\text{-}35\text{a})$$

$$D(s)/E(s) = C + I(s) = C + \Sigma\, K_i/(s + k_i) \qquad (8\text{-}35\text{b})$$

Then one generates $2N + 1$ different $D(s)/E(s)$'s and solves these equations to obtain τ's, K's, and C. Gafni *et al.* (1975) found that, even with substantial stray light ($C = 0.7$), successful deconvolution and estimates of C could be made for a two-exponential system.

The treatment for a time shift is equally straightforward, and the reader is referred to the original paper for a derivation. Experimentally, Gafni *et al.* (1975) were able to use this shift correction to measure the single-exponential lifetime of 9-cyanoanthracene. With their particular PMT the shift was 0.27 nsec between the 340-nm excitation and 450-nm emission wavelengths. Once the time shift is available for given experimental conditions, it can be used to correct other data for time shift errors.

Closed form solutions for a number of cases are presented in the original article by Gafni *et al.* (1975). These authors also include a detailed discussion of the selection of the range of s's to use. This article should be read carefully by anyone wishing to implement the Laplace method.

Several points should be made about the Laplace method. It is very flexible in terms of the types of errors it can correct for (e.g., time shift and scatter). Other impulse functions are easily incorporated as long as one can write down the Laplace transform. In computational complexity it is probably between the moments methods and nonlinear least squares method. It lacks statistical information inherent in other methods.

In spite of its successful utilization by Gafni *et al.* (1975) there are questions about its use. In a comprehensive test of a number of different methods McKinnon *et al.* (1977) found the Laplace method to fail with their experimental data. The apparent discrepancy between the Gafni *et al.* and the McKinnon *et al.* work can probably be traced to the choice of s values. O'Connor *et al.* (1979) have shown that for noisy real data Eq. (8-33) holds only over a limited range of s. Thus, if the wrong range of s is inadvertently selected, severe errors can result. O'Connor *et al.* have also extended the Laplace method so that limited regions of the decay can be fit. These results show that considerable caution is necessary in proper use of the Laplace method, and experience can be a significant factor in its success.

We turn now to Fourier transforms. The history of Fourier transforms in deconvolution has been long and generally unsuccessful. It now appears that the Fourier method will fill a valuable niche in deconvolution techniques (O'Connor et al., 1979; Andre et al., 1979).

Fourier transforms of $E(t)$ and $D(t)$ to the frequency domain are given by

$$D(\nu) = \int_{-\infty}^{+\infty} D(t) \exp(-2\pi i \nu t) \, dt$$
$$E(\nu) = \int_{-\infty}^{+\infty} E(t) \exp(-2\pi i \nu t) \, dt$$

(8-36)

The sample impulse response $i(t)$ is then given in the frequency domain by

$$i(\nu) = D(\nu)/E(\nu) \qquad (8-37)$$

This expression is derived similarly to the analogous one for Laplace transforms. Now, in principle, a simple inverse Fourier transform of $i(\nu)$ back to the time domain should yield $i(t)$. Unlike the Laplace transforms the necessary inverse Fourier transforms are easily carried out. In practice, however, the noise generated by the division in Eq. (8-37) in the frequency domain has caused $i(t)$ to be totally worthless. Numerous attempts to alleviate this problem using various filters and frequency cutoffs in the frequency domain have, until recently, been unsuccessful (see, e.g., McKinnon et al., 1977). We briefly describe recent developments that promise a more satisfactory resolution.

Wild et al. (1977) have demonstrated that, if a functional form for $i(t)$ is assumed, then it is possible to perform the least squares fitting directly in the frequency domain. Further, these authors have shown how the necessary weighting factors in the frequency domain can be derived from the single-photon counting statistics in the time domain. While currently applied only to very fast single-exponential impulse responses, the method seems, in principle, extendable to more complex decay schemes.

Andre et al. (1979) have published a detailed paper describing a frequency cutoff approach. Normally, ignoring the high-frequency region causes the back-transformed $i(t)$ to oscillate so wildly as to be useless. Andre et al., instead of ignoring this region completely, fill it in by assuming an exponential form and extrapolating into the high-frequency region. The approach seems to work well because most functions approach exponential behavior. Now sufficient data are available to take the inverse transform and generate $i(t)$. The results are much less oscillatory than those obtained by previous techniques, and the generated $i(t)$'s are satisfactorily fit by the standard nonlinear least squares method. We refer the reader to the original paper for full details and results of the method.

The Andre Fourier method has many appealing features. It permits direct generation of $i(t)$ that can be fit by any model. It is much less sensitive to noise than previous Fourier methods. It has several disadvantages. The $i(t)$'s generated are still oscillatory. Particularly if slopes of $i(t)$ are desired, smoothing of the functions is required. Finally, the computational requirements of the method are rather high.

G. LEAST SQUARES FITTING

Least squares fitting is one of the most popular approaches to deconvolution. It is also known as deconvolution by iterative reconvolution. As in Chapter 5, one merely varies the parameters of the modeling equation until the χ^2 calculated for the model is a minimum. The simplex, analytical, Marquardt, and other methods mentioned in Chapter 5 are suitable for determining the fit. In practice, the calculations, especially with the fitting of a sum of exponentials, are quite formidable, and a fast algorithm such as the Marquardt is preferred. We describe briefly the implementation of these methods.

For an impulse response equal to a single exponential the simplest of grid searches is reasonably efficient for deconvolution, and it is even possible to reduce the search to a single parameter. One merely guesses a lifetime τ_{guess}, calculates the expected decay $D^{calc}(t)$, and compares it with the observed $D(t)$. To avoid having to fit the preexponential factor, one can normalize $D^{calc}(t)$ either to the same peak height as $D(t)$ or to the same area as under $D(t)$ and then calculate χ^2. Area normalization is more accurate, particularly for analog data, because peak normalization puts excessive weight on a single potentially noisy point, whereas the area approach averages over the entire curve. For reasonably noise-free data, however, the two approaches are virtually identical. One then goes back and continues the variation of τ_{guess} to minimize χ^2 using a simple one-dimensional grid search.

A better approach in a single-exponential least squares deconvolution is to minimize χ^2 with respect to both the lifetime and the preexponential factor. A two-parameter grid search is suitable, but the simplex method is more efficient.

If the assumed impulse is a sum of exponentials, the calculations can become unacceptably slow with simple fitting methods. The analytical and Marquardt methods can be used to accelerate convergence. A problem immediately arises, however. These methods require the $\partial \chi^2/\partial a_j$ for all of the a_j parameters. There are two ways to evaluate these partial derivatives, numerically and analytically. The numerical one is based on the definition of a partial derivative

$$(\partial \chi^2/\partial a_j) = \lim_{\delta a_j \to 0} [\chi^2(a_j + \delta a_j) - \chi^2(a_j)]/\delta a_j \qquad (8\text{-}38)$$

where $\chi^2(a_j + \delta a_j)$ and $\chi^2(a_j)$ correspond to χ^2 evaluated at $a_j + \delta a_j$ and a_j, respectively, with all the remaining a_i $(i \neq j)$ held constant. For small enough a_j, the right-hand side approaches the required partial derivative acceptably closely. A disadvantage of this approach is that, for an n-parameter system, χ^2 must be evaluated n additional times to obtain the necessary partial derivatives. Further, there can be difficulty in evaluating the derivatives. If too large a δa_j is chosen, the derivative becomes inaccurate or even meaningless (e.g., if the two χ^2's are evaluated on either side of a ridge or valley). This problem can be minimized by evaluating χ^2 at $a_j - \delta a_j$ rather than a_j for the difference (Bevington, 1969), but n more χ^2's are required. If δa_j is too small, the difference between the two terms can be lost in the resolution of the computer arithmetic. It is, therefore, common to use double-precision arithmetic when evaluating such derivatives numerically.

For the common case of an impulse response that is a sum of exponentials it is possible to evaluate the necessary derivatives analytically. Probably the most popular approach originated with Grinvald and Steinberg (1974), although Zimmerman et al. (1974) independently presented a similar treatment which is also widely used. We confine our discussions to the Grinvald–Steinberg method.

Grinvald and Steinberg (1974) used an extension of their fast iterative convolution integral [Eq. (6-20)]. One assumes that Eq. (6-20) is exact and directly evaluates

$$(\partial D_{j+1}/\partial \tau) = P[(D_j + SE)(\Delta t/\tau^2) + (\partial D_j/\partial \tau)] \qquad (8\text{-}39)$$

$$P = \exp(-\Delta t/\tau), \qquad S = K \, \Delta t/2$$

$\partial D_{j+1}/\partial K$ is obtained directly from the convolution integral:

$$\partial D_{j+1}/\partial K = \frac{\partial}{\partial K} \sum_{i=1}^{j} E_i K \exp[-(j - 1) \, \Delta t/\tau]$$

$$= E_j K \qquad (8\text{-}40)$$

As with the fast iterative convolution integral for evaluating D_{i+1}, the iterative expressions are far faster than direct numerical evaluation of the derivatives. For a sum of n exponentials there will be n different equations of the form of Eqs. (8-39) and (8-40). We can now apply the generalized expressions of Chapter 5. Figure 8-6 shows typical single-photon counting data fit by least squares.

Extensive tests have been carried out on the nonlinear least squares deconvolution procedure by Grinvald and Steinberg (1974), McKinnon *et al.* (1977), O'Conner *et al.* (1979), and Love and Shaver (1980). We summarize only a few results, and the interested reader should consult these references for details. The nonlinear least squares approach is a reliable, versatile method of determining sample parameters for simple and com-

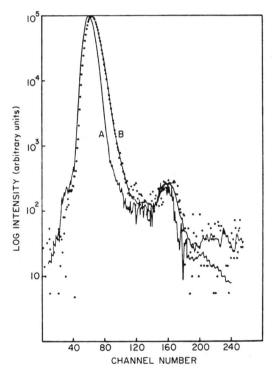

Fig. 8-6. Nonlinear least squares fit of the single-exponential decay of POPOP in cyclohexane. Each channel corresponds to 0.25 nsec. The solid curve A is $E(t)$ and B is $D(t)$ with the dots being the experimental points and the solid curve being the calculated fit. The short lifetime of the sample (1.1 nsec) relative to 3.5 nsec width (FWHM) of the flash makes a demanding deconvolution problem. [Reprinted with permission from O'Connor *et al.* (1979). Copyright 1979 American Chemical Society.]

plex decay kinetics. It can correct for time shift and scattered light (Grinvald, 1976).

The nonlinear least squares approach is capable of reliably deconvoluting a sum of two exponentials when the lifetimes are quite close. For example, Grinvald and Steinberg (1974) carried out simulated deconvolutions using a 10-nsec flash (FWHM) and biexponential decays with $\tau_1 = 5$ nsec, $\tau_2 = 7$ nsec, and $K_1/K_2 = 1$. They found that, with Poisson noise and weighted least squares, τ_2, τ_1, and K_1/K_2 were accurate to 7.5, 6.5, and 68%, respectively. When the data were unweighted, however, the errors increased to 14, 15, and 146%, respectively.

O'Conner et al. (1979) have had excellent success with the least squares method even under conditions where there were nonspecific distortions of the leading edges of the decays. They just performed the least squares fitting to the later undistorted portions of the decays. Such a procedure, of course, sacrifices some information inherent in the signal.

Under favorable circumstances a sum of three exponentials can be deconvoluted. Resolution of two or three lifetimes does, however, necessitate very high-quality data such as those arising from single-photon counting (O'Connor et al., 1979) or a refined signal-averaging boxcar system (Grinvald and Steinberg, 1974).

Love and Shaver (1980) also tested, by simulations of single-photon counting experiments, the suitability of the least squares method for the determination of short lifetimes. Using a 3-nsec-wide flash (FWHM) with a 1.3-nsec decaying tail, they were able to deconvolute lifetimes of 0.05 nsec with 10% accuracy. These theoretical results should be compared with the experimental results of Lyke and Ware (1977) on neat deoxygenated n-hexane. The experimental lifetime was 155 ± 60 psec, which demonstrates that the instrumentation still falls somewhat below the theoretical performance.

Least squares are particularly useful because they are easily adaptable to any decay model. Further, they utilize the maximum amount of statistical information in data and yield the maximum amount of statistical information about the final results.

The principal problem with deconvolution by the nonlinear least squares method is the high computational demand. While it is dangerous to generalize about the speed of mathematical algorithms, O'Connor et al. (1979) suggested that nonlinear least squares and Fourier transforms were comparable but typically two times slower than most other approaches. Not surprisingly, however, speed depends on the particular data, and in at least one case the moments method was twice as slow.

We conclude with comments on the judging of fitting. As discussed in Chapter 5, common methods of judging the suitability of a model are the

χ^2 test, autocorrelation plots, and weighted or unweighted residual plots. All these methods work quite satisfactorily and are widely used. Weighted residual plots are preferred for single-photon counting data where the statistics are known, whereas autocorrelation is preferable for sampling instruments or transient recorder data where statistical information is not available. We offer one word of caution, however. These methods all work well if there are a large number of relatively noise-free uncorrelated data points for $E(t)$. High noise levels in $E(t)$ can, however, propagate or mask correlated errors in $D(t)$, which yields both poor χ^2's and correlated drift in the residual plots (Irvin *et al.*, 1981). In general, with high-quality data these effects will not be observed, but with noisy data as might be obtained in flash photolysis, with streak cameras, or with some analog detection systems, their possible appearance should be recognized. Irvin *et al.* describe a complex generalized statistical χ^2 test suitable for such data.

H. COMPARISON OF DECONVOLUTION METHODS

Table 8-3 compares important features of some of the deconvolution methods. A variety of factors go into the choice of a method. There is, for example, almost always an element of personal preference or prejudice in such a selection, and the ready availability of programs and workers versed in the particular techniques can have a decisive effect on the choice.

TABLE 8-3

Comparison of Deconvolution Methods

	Moments	PP	Laplace	Fourier	Nonlinear least squares
Multiple-exponential fits	Yes	No	Yes	Yes[a]	Yes
Other decay function fits	Limited	Yes	Yes	Yes[a]	Yes
Complexity	Simple	Simple	Moderate	High	High
Speed	Fast	Very fast	Moderate	Slow	Slow
Stray light corrections	Yes	Yes	Yes	No	Yes
Time shift corrections	Yes	No	Yes	No	Yes
Fit to any region of decay curve	No	Yes	No	Yes[a]	Yes
Cutoff corrections required	Yes	No	Yes	Yes	No
Statistical uncertainties	No	Yes	No	Yes	Yes
Runs noninteractively	No	Yes	No	No	Yes

[a] Impulse response of sample is obtained, which in turn must be fit to decay functions.

The initial question is whether one will be deconvoluting single- or multiple-exponential impulse responses or some more complex function. For single or double exponentials all methods are suitable. For other types of decay functions, only nonlinear least squares and the Fourier method are perfectly acceptable. The phase plane and moments methods have, however, been extended to Förster or diffusional transients. The phase plane method also works with a single-exponential decay and a constant but unknown baseline (Bacon and Demas, 1983). It seems likely that other decay functions could be fit by the Laplace and moments methods. For most workers, however, a sum of exponentials is usually adequate.

The term *complexity* refers to programming complexity. The nonlinear least squares method, and especially the Fourier method, have complex programming algorithms, although frequently computer system library routines can handle most of this difficulty. The phase plane method is the simplest and can easily be run on a programmable hand-held calculator. The moments methods can be appreciably more complex, especially for multiple-exponential fits. For a single-exponential fit they can be comfortably run on a hand-held calculator.

"Speed" refers to the computational speed of deconvolution. For computations done on large mainframe computers, all methods are probably acceptable. In a mini- or microcomputer environment, however, speed can become a limiting factor. For a single-exponential deconvolution, the phase plane method has no equal in speed. For micro- and minicomputers speed should be acceptable for a single-exponential fit by all methods. For multiple-exponential evaluations the moments method is probably the fastest, and the nonlinear least squares and the Fourier transforms approaches are probably the slowest (O'Connor et al., 1979). Fourier transforms and nonlinear least squares methods are both slow enough to be bothersome, especially when computations are performed on a microcomputer.

Several important points such as the ability to correct for stray light and time shift can be critical. The Fourier method does not account directly for either of these. The ability to fit only a limited portion of the curve can also be important. Only the nonlinear least squares, Laplace transforms, and phase plane methods can be used. In this case it should be recognized that all flash and decay data from $t = 0$ through the region being fit are actually being used. One merely has the option of ignoring these regions where the pertubations are most severe.

Methods that use integrals to infinite time generally have stringent requirements on data quality at long times, although the Laplace method minimizes the problem by its inherent exponential depression. In addition, these techniques may require operator intervention so as to ensure convergence or optimization of the answers. The use of exponential depression in the moments method or exponential expansion in the Laplace method exemplifies these features. In particular, the need for the operator to adjust these parameters manually and never having clear criteria for selection of the best parameters are the primary disadvantages of the moments methods. Also, when the operator must interact during the course of the calculation, much of the speed and pristine elegance of the moments method is lost.

In summary, it seems likely that, for workers who wish the minimum interaction with their calculations and the maximum amount of statistical information, the nonlinear least squares approach will be the method of choice. Where speed and programming simplicity at the expense of operational complexity are the criteria, the moments method will continue to attract a staunch group of supporters. Laplace transforms will be used only occasionally for specialized problems such as stray light and time shift corrections. For single-exponential decays, the phase plane method is unequaled in speed and simplicity.

Finally, we think one important point should be noted. The speed of convergence of nonlinear least squares is determined in part by the guess. The moments method frequently can be used to give excellent initial guesses, which permits much more rapid convergence of the nonlinear least squares while still providing all the statistical information.

We add that there are substantial philosophical differences among many researchers. In one view true deconvolution without an assumed model is considered to be unreliable and to create the agreeable impression that new information has actually been extracted. The only acceptable procedure is to assume a model and evaluate the best-fit parameters (Knight and Selinger, 1971). Others feel that the best procedure is, if possible, deconvolution to $i(t)$ and modeling directly on $i(t)$. If used with common sense and without pushing data or data reduction beyond its limits, both viewpoints will generally yield the same conclusions.

Experimental Methods

A. INTRODUCTION

In this chapter we supply further information on experimental methods. As with all other areas of excited state lifetime measurements, the subject is far too extensive to do justice to here. We, therefore, supply key references where available, indicate sources of further and future information, add a few methods not previously discussed, and comment on potentially useful further developments.

Everyone interested in experimental methods should read the classic paper by Ware (1971). This article not only summarizes much of the technology available up to 1971 but in a lucid fashion gives detailed experimental methods for carrying out lifetime measurements. West's (1976) article on flash photolysis is another exceptional article and should be read by anyone interested in the subject.

Another excellent review article is by Imhof and Read (1977). These authors discuss extensively lifetime measurements with special emphasis on measurements on atoms, molecules, and ions in the gas phase and in beams. While many of the approaches are the same as those described here, a number of the coincidence techniques are unique. The original article should be consulted.

For a single source of information on optical detectors, lasers, optics, and modulators, the latest issue of *Laser Focus Buyers' Guide* is without equal. The 1982 issue consists of 608 pages full of ads, general information, and a subject index listing suppliers' and manufacturers' addresses and phone numbers. It is an indispensable addition to any library.

In terms of general information on new products, the trade magazines are one of the best ways to learn of new developments. In particular, for optics and optically related electronics, lasers, and detectors, *Laser Focus, Optical Spectra,* and *Electro-Optical Systems Design* are excellent sources of up-to-date information. For electronics and computers, *Electronics, EDN,* and *Electronics Design* cover most new developments. Hobbyist magazines are also an excellent way to learn of new microcomputer software and hardware developments; *Byte, Microcomputing,* and *Doctor Dobb's Journal* are good general sources.

B. LIGHT SOURCES

Conventional pulsed light sources have improved remarkably little since Ware reviewed them in 1971. Munro (1983) has summarized the properties and applications to lifetime measurements of synchrotron sources; the high cost and limited availability restrict their use, and we will not discuss them further here. Laser sources, however, have replaced conventional light sources in many areas. The wealth of commercially available lasers is exemplified by a listing of more than 20 pages in the 1982 *Laser Focus Buyers' Guide.* We summarize only some relevant characteristics of important systems. Potential users should consult the manufacturers for further details.

For fast flash photolysis, ruby lasers (fundamental, doubled, and tripled) and 1060-nm neodymium lasers (doubled, tripled, and quadrupled) can typically supply 0.028–>1 J in the visible to uv regions. Pulse widths of 10–20 nsec are typical, and picosecond pulses are possible. The cost of these systems is high.

Pulsed flash lamp-pumped dye lasers can provide >1–3 J of energy in 1 μsec; 0.1-μsec pulses are available at lower energies. They are tunable from 350 to >700 nm. They can supply useful powers when frequency-doubled to ~220 nm. For many flash photolysis applications, the long pulse width is not a disadvantage, and the cost of flash lamp-pumped dye lasers is much less than that of solid state lasers.

Rare gas excimer lasers can yield remarkably large amounts of energy in short pulses (2–20 nsec) in the UV at modest costs. Representative systems are ArF (193 nm, 200 mJ), KrCl (220 nm, 50 mJ), KrF (248 nm, 1 J), XeBr (282 nm, 10 mJ), XeCl (308 nm, 10 mJ), and XeF (350 nm, 0.2 J).

Indicated wavelengths and powers are the operating wavelengths and representative single-shot output energies. They can be used to pump dye lasers.

Nitrogen lasers are of relatively modest cost and can be fabricated easily and inexpensively; see the references given by Demas (1976). Phase R sells a system for <$2000, but EMI problems can be serious (Matthews and Lytle, 1979). Most N_2 lasers exhibit ~10-nsec widths with powers of 0.1–5 MW. Atmospheric pressure units yield subnanosecond lifetimes. Commercial lasers have 0.3-nsec pulse width and peak powers of 100-kW or more.

A particularly nice application of a nitrogen laser–pumped dye laser system was presented by Hammond (1979). He measured the fluorescence decay times and self-quenching of several laser dyes. He shortened the dye pulse to ~1 nsec using a saturable absorber. He measured the decays with a fast vacuum photodiode and a conventional fast oscilloscope—a combination that would not be considered useful by many. Figure 9-1 presents some of his data.

Rare gas, nitrogen, and solid state lasers can be used to pump dye lasers, which extends the range of operation. See, for example, the comprehensive article by Wallenstein (1979) on dye lasers.

The subject of picosecond pulse generation is beyond the scope of this book. The interested reader is referred to the articles by Shank and Ippen (1977) and Lowdermilk (1979). The brochure by Spectra-Physics (1981) has a very clear elementary discussion. The most popular technique is mode locking in which a pulse train is formed in the cavity and sweeps out the excited state concentration on each pass. Thus, each pulse is sepa-

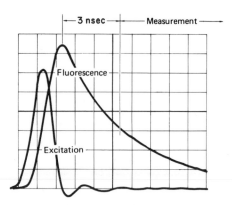

Fig. 9-1. Excitation and decay transients for 10^{-7} M rhodamine 6G dye recorded on an oscilloscope using a photodiode detector. Each horizontal division is 1 nsec. [From Hammond (1979).]

rated by the round-trip time in the cavity which is $2L/c$, where c is the speed of light in the laser medium. Generally, pulse widths are ~100 psec, peak powers are in the hundreds of watts, and frequencies are ~100 MHz. In cavity dumping (Harris et al., 1975) the energy is stored and dumped out less frequently but with considerably greater power per pulse (~30 times) (Spectra-Physics, 1981). The loss of repetition rate is generally not serious, as most decay time systems cannot exploit the high rate.

In summary, flash lamps maintain their position in low-speed flash photolysis instruments and in relatively inexpensive single-photon or sampling systems. Where a shorter time resolution, especially below 1 nsec, is required, lasers are the excitation sources of choice.

C. OPTICAL DETECTORS

The primary detectors are PMTs and photodiodes. The extensive *Photomultiplier Handbook* (RCA, 1980) gives an excellent discussion of PMTs and is a must for anyone's library; see also the articles by Lytle (1974). Again, the latest *Laser Focus Buyers' Guide* is an excellent source of current suppliers and information. The important recent advances in photomultipliers are principally in the area of speed. Static cross-field and microchannel plate PMTs offer rise times of 100–300 psec and have sufficient gains to be used with modern single-photon counting equipment. Mura et al. (1982) have reported an SPC laser lifetime system with a microchannel PMT which exhibited a 73-psec (FWHM) impulse response.

We should add that, with advances in low-level discriminator design, the lowly and once much maligned side-view squirrel cage PMTs have once more come into their own. They can supply sufficient gain to be used without a preamplifier in a single-photon counting instrument (O'Connor et al., 1979; Kinoshita et al., 1981; Kinoshita and Kushida, 1982). Further, because of the compact configurations, they yield remarkably quick rise times compared to much more expensive, fast, end-on tubes. For example, Harris et al. (1976), using a special dynode string, have measured rise times of 1.35 nsec and widths (FWHM) of 2.3 nsec for an inexpensive 1P28 photomultiplier. They found, however, that these tubes exhibited a position-dependent transit time which varied by a maximum of 1 nsec (see also Kinoshita et al., 1981; Kinoshita and Kushida, 1982). This result stresses again the need for very uniform or very limited illumination of the photocathode.

Streak cameras with down to a 2-psec resolution currently hold the speed record for direct recording of fast optical phenomena. Their exceptionally high cost and limited dynamic range, however, limit their widespread usefulness.

D. OPTICAL MODULATORS

The *Laser Focus Buyers' Guide* provides an extensive compilation of modulators and drivers. They are generally divided into electrooptical and acoustooptical devices.

Electrooptical devices usually have perpendicular entering and exiting polarizers; the electrooptical device can rotate the polarization of the beam as it passes through. If the beam is not rotated, the crossed polarizers block it. If the rotation is 90°, the light is transmitted. Pockel and Kerr cells fall into this class. They can have subnanosecond rise times and modulate signals to 1 GHz. They generally require large operating voltages, and the extinction ratio can be low (2% transmittance for the very fast ones). Further, those most suitable for very high-frequency modulation have small apertures and acceptance angles and are, thus, suitable only for laser sources. Electrooptical modulators are ideal modulators for laser-based phase shift instruments.

In acoustooptic modulators a high-frequency acoustic modulation is applied to a clear block. Once a standing wave is established, a periodic diffraction grating is set up from the variations in the refractive index of the medium. When the rf drive is off, light is transmitted directly through the block. When the rf excitation is on, the beam is diffracted by a few degrees. The diffracted beam can contain >90% of the beam energy, whereas the off state is essentially perfect. The switching speed is slower than for electrooptical devices, since establishment of the standing wave is determined by the velocity of sound in the medium. Modulation frequencies are typically 1–10 MHz, although some will work to 100 MHz and have 4-nsec rise times. The drive circuits for the modulation of acoustooptic devices use lower voltages and are generally simpler and less expensive than for electrooptical devices.

E. SINGLE-PHOTON COUNTING INSTRUMENTATION

There are many articles on do-it-yourself single-photon counting instruments (Ware, 1971, 1983; Cramer and Spears, 1978; Spears *et al.,* 1978; Cantor and Tao, 1971; Wild *et al.,* 1977). The key to success for all but those extremely sophisticated in electronics is to buy critical components such as the TAC, the discriminators, and the multichannel analyzer, from reputable manufacturers of single-photon counting or nuclear equipment. Very inexpensive ($150) pulse amplifier discriminators have, however, been described by Borders *et al.* (1980) and by Kinoshita *et al.* (1981). Many build their own flash lamps (Ware, 1971), but the serious problems of EMI (Chapters 10) have frequently led to failure of such projects. Therefore, purchase of a suitable flash lamp is also recommended if funds permit. Researchers will, however, find manufacturers extremely helpful in assembling a minimal system and in providing technical assistance.

The major advances in instrumentation have been the introduction of constant-fraction discriminators (CFDs) and mode-locked lasers. CFDs avoid timing jitter associated with different rise times for the photoelectron pulses; for a detailed discussion of the operation see the ORTEC (1975) Model 473 CFD manual. Mode-locked lasers exhibit 100-psec pulse widths and are discussed in greater detail in Section B.

Cramer and Spears (1978) have discussed a complete single-photon counting instrument based on a mode-locked laser. Noteworthy features are the use of a silicon photodiode for generating the start pulse (the higher laser intensities avoid the need for a PMT), and the 0.8-nsec-wide instrument response function. More recently Spears *et al.* (1980) improved this to 0.29 nsec. This impulse response is caused by PMT and electronics jitter and not by the mode-locked laser which probably has a pulse width of 100 psec. The approximately 100-MHz pulse rate could not, however, be utilized because of the speed of the PMT and MCA. Typical acceptance rates were $1-6 \times 10^4$ Hz, which permitted acquisition of a complete decay curve with 20,000 counts in the peak channel in only 0.5–3 min. This is an enormous improvement over conventional flash instruments where the count collection rate might be only 100–5000 Hz.

Detector resolution, however, can be improved on. Koester (1979) has used a static crossed-field photomultiplier. This tube has a response pulse width (FWHM) of 160 psec, and in a single-photon counting instrument he observed a 228-psec-wide (FWHM) impulse response. A similar performance has been obtained with side-view tubes (Kinoshita *et al.*, 1981). Microchannel plate PMTs yield comparable or superior performances (Zimmerman *et al.*, 1982; Mura *et al.*, 1982).

F. PHASE SHIFT METHODS

Spencer (1970) and Lakowicz (1983) have reviewed the phase shift method. Spencer (1970) has a lucid discussion of the error sources as well as complete plans for a phase shift lifetime instrument. There have been several recent noteworthy advances. Light-emitting diodes (LEDs), although limited in their wavelength range (500–1000 nm), make excellent sources; Moya (1983) has presented beautiful data on a photosynthetic system using these diodes. They can be modulated at >80 MHz, and advances in fiber optics communications hardware promise usable systems to perhaps 1 GHz. Of course, because they can be modulated at any lower frequency, continuous tunability of frequency and far greater information in phase shift measurement are available.

Using a cw laser excitation source with Pockel cell modulation, Haar and Hauser (1978) have constructed a phase fluorometer with a 5–500-MHz operating frequency. This system is capable of resolving less than 1 psec, and the authors have used it to measure rotation times as short as 25 ± 1 psec (Haar et al., 1977).

Schubert et al. (1980) have used a mode-locked He–Ne laser (78 MHz) as the modulated light source. An optical delay line was used for measuring the phase angle. The system has a claimed resolution of ±30 psec. The principal disadvantage is the fixed frequency imposed by the laser's mode lock characteristics.

We have not mentioned at all the very elegant and powerful differential phase fluorimetry. In this approach, the sample is excited with polarized light and the phase difference between the vertical and horizontal polarized emission is measured. This contrasts with the normal phase shift method where the shift is measured between the excitation and emission. Differential phase fluorimetry is particularly useful for studying the rotational characteristics of small molecules and proteins (Mantulin and Weber, 1976; Lakowicz, 1981, 1983). These measurements would be difficult and time-consuming with single-photon counting instruments.

The use of phase fluorimetry for time-resolved spectroscopy is discussed in Section G. Again, phase fluorimetry is superior in convenience and speed to single-photon counting for short-lived samples where the sample impulse response is a single-exponential decay.

G. TIME-RESOLVED SPECTROSCOPY

There are three common ways to measure time-resolved emission spectra: multiple-transient recording, boxcar approaches, and phase shift methods. We describe each briefly.

In the transient recording method, one records a complete decay curve at each emission wavelength. If necessary, deconvolution to obtain the sample's impulse response can be performed on each. Then, one merely picks a delay time and plots the intensity at this delay for the decay at each emission wavelength (Easter *et al.*, 1976). Such a procedure can be especially laborious and time-consuming if a standard single-photon counting instrument is used. Aoki and Sakurai (1980) have reported an instrument that uses single-photon counting but divides the decay period up into 32 equal time periods (minimum of 25 nsec/window) and counts and stores the number of photons in each of these windows on each occurrence of the excitation pulse. In this way, at some sacrifice of temporal resolution, many photoelectrons can be accumulated per pulse rather than 0.01–0.10 per flash in a normal single-photon counting instrument. For such a design long lifetimes are required.

Boxcar methods work just like the standard boxcar integrator described in Chapter 2, except that the delay time between the trigger and the sampling pulse is held fixed rather than varied. The emission spectrum is scanned while monitoring the output of the boxcar integrator. In this way one directly determines an emission spectrum. For a beautiful example of time-resolved spectra taken using a sampling approach, see the paper by Chakrabarti and Ware (1971) or, more recently, the work of Choi *et al.* (1980) which was done on a subnanosecond resolution decay time instrument (Badea and Georghiou, 1976).

An elegant approach uses phase-resolved luminescence spectra. It is suitable where the components of a mixture are characterized by simple exponential decays, which is common in dilute solutions. It is like a conventional phase shift measurement except that one takes advantage of the fact that the total luminescence is a sum of the sinusoidal emissions of all the components. Each, however, has a phase shift characterized by its lifetime. A phase-sensitive detector operating at the modulating frequency is used. The detector's phase angle is then adjusted so that the detection is 90° out of phase with the signal of one of the components. The phase-sensitive detector then completely ignores that component and can see only samples with other phases. The process can be repeated with the detector's phase shift adjusted to eliminate the other component. In this manner, the essentially undistorted luminescence spectrum from each of the components can be obtained. This approach was originally described by Veselova and co-workers (Veselova *et al.*, 1970; Veselova and Shirokova, 1972) and has been refined and described in detail by Lakowicz and Cherek (1981a,b) and by Lakowicz (1983). It is now used in a commercial instrument (Mitchell and Spencer, 1981). Figure 9-2 shows results obtained with the system. In this case the two components have lifetimes of 1.0 and 1.2 nsec and are extensively overlapped. Resolution of such short, close lifetimes in time-resolved spectroscopy by any other approach would be difficult. Phase-resolved spectra have the advantages

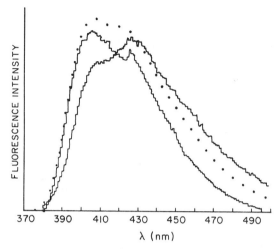

Fig. 9-2. Phase-resolved spectrum of POPOP and dimethyl-POPOP in methanol. The dotted line is the total emission spectrum. The solid lines are the phase-resolved spectra of each component with the high-energy emission from POPOP and the low-energy one from dimethyl-POPOP. $\lambda_{ex} = 355$ nm; $\Delta\lambda_{ex} = 1$ nm; $\Delta\lambda_{em} = 8$ nm; $f = 30$ MHz; 10^{-6} M concentration. [From Mitchell and Spencer (1981).]

of a short acquisition time and superb temporal resolution; the 30-MHz excitation of Fig. 9-2 is low for optimum resolution of the two components. Phase-resolved spectra do not, however, directly give relative intensity information about the two components. Further, this method is not currently suitable for the complex spectra of solvent relaxation systems (DeToma, 1983) or for systems with more than two components.

In summary, the three approaches should be considered complementary rather than competitive. Depending on the system and the information desired, each approach can supply valuable information faster or with a temporal resolution better than that of the others. Overall, the single-photon counting approach is the most sensitive and versatile, but in speed and convenience it is inferior to the others.

H. TRANSIENT RECORDERS AND SIGNAL AVERAGERS

A transient recorder (TR) or digital oscilloscope has a built-in clock time base, an analog-to-digital converter (ADC), and a memory. It digitizes and stores in memory a waveform at a user-specified rate. Once stored, the data can be displayed repetitively on an oscilloscope screen for the user's leisurely inspection. Many TRs have provisions for slow output of the waveform so that it can be displayed on an analog recorder; also, some TRs have interfaces for transferring data to computers. Generally, the distinction between digital oscilloscopes and TRs is that the former have built-in oscilloscopes and the latter do not. Commercial transient recorders (Santoni, 1980) and digital storage oscilloscopes (Nelson, 1981) have both been reviewed. A PMT operated in the single-photon counting mode can also be interfaced to an MCA to yield a signal averager (Donnelly and Kaufman, 1977).

The cost and performance of transient recorders have improved enormously over the past few years. For relatively low-speed phenomena (2–10 data points/μsec), where significant amounts of data reduction are required, it no longer makes good sense to use oscilloscopes, photographic data recording, hand digitization, and manual entry of data into a computer. A relatively inexpensive transient recorder equipped with a recorder output will provide higher-quality data than an oscilloscope and reduce enormously the digitization problem. Far and away the best procedure, however, is a system with a computer interface. Data can then be directly transferred to the computer and reduced without having to go through film or recorder paper. It is, thus, possible to acquire a decay and have the kinetic information within less than a minute, which permits more and better experiments to be performed and frees the operator from the drudgery of mechanical data reduction.

We touch briefly on the character of transient recorders. Most such devices use standard analog-to-digital converters. Typically, 8–10-bit data can be acquired at 2–10 points/μsec. Typical storage capabilities are 1024–4096 points per transient, with some instruments organized to permit storage of one, two, or four transients. The Biomation 8100 has 8-bit resolution with a maximum 100-MHz digitization rate (10 nsec/point), and the Biomation 6500 exhibits 6-bit resolution with a 500-MHz digitization rate.

Currently, the speed king is the Tektronix 7912 series transient digitizer which uses a completely different technology. A double-ended scan tube is used with a 0.38 × 0.5 in. diode array with ~2000 diodes per axis. One end of the tube is a very high-speed oscilloscope which writes on the array, and the other end is a scan gun which interrogates the array to determine where the electron beam strikes the target. The data can be digitized during the read cycle. The newest versions have a built-in microprocessor which converts the very complicated target readout into a single-valued waveform for display or transfer to a computer. Earlier versions supplied the data from the target in the raw digital form, and it was the user's responsibility to convert it to a single-valued waveform—a not insignificant task (Turley, 1980; Tektronix, 1976). By going directly to the converter tube, the bandwidth is 2 GHz. With vertical plug-ins, however, reasonable sensitivity is obtained with a rise time of 0.6 nsec (500 MHz). Effective resolution is 8–9 bits with 512 points on the vertical axis. Figure 3-1 was taken on a 7912 digitizer which demonstrates the data quality.

There are also a number of dedicated signal averagers on the market. For dedicated applications where the highest averaging rate is essential, these devices represent a viable approach. In the author's opinion, however, most users will be much more satisfied with a transient recorder equipped with a computer interface and a small dedicated microcomputer. Data can be transferred to, and averaged by, the computer after the acquisition of each transient. This also provides complete on-line data reduction and display capabilities. As an added benefit, one can expand, examine, and manipulate a small portion of the memory, which cannot be done on most transient recorders. Thus, one does not have to pay for the memory and averaging circuitry in the averager. For example, the flash photolysis system of B. A. DeGraff (private communication) uses a 10-bit

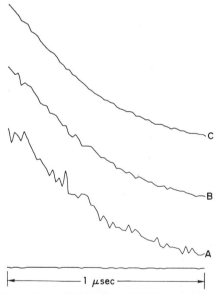

Fig. 9-3. Nitrogen laser (12 nsec FWHM)-excited luminescence decay curve of tris(2,2′-bipyridine)ruthenium(II) in air-saturated water. A, Noisy single-sweep decay; B, decay after 32 averages; C, decay after 128 averages. The baseline is correct for A, but B and C have been displaced for clarity. The position of the baseline for B and C is indicated by the dots on the right-hand side of the figure. The data were taken on the microcomputer-controlled boxcar integrator described by Taylor *et al.* (1980).

2-MHz Physical Data 514A transient recorder with an IEEE-488 interface, a Hewlett-Packard 85A computer with full CRT graphics and mass storage capabilities, and a publication-quality Hewlett-Packard 7225A digital plotter. Signal-averaging and all data reduction including nonlinear least squares fitting can be performed with this system in a reasonable amount of time. The total cost of the system is <$8000.

There are also a number of very nice averaging schemes based on sampling oscilloscopes, especially when interfaced to computers. Badea and Georghiou (1976) have described a complete analog system for subnanosecond lifetime measurements. Plumb *et al.* (1977) described an LSI-11 microcomputer-based transient recorder signal averager with an 0.35-nsec rise time. Taylor *et al.* (1980) described a similar 0.35-nsec rise time 8-bit microcomputer-based system; their system permitted signal averaging, baseline subtraction, oscilloscope or $X-Y$ recorder display of results, and a capability for transferring data to a larger computer. The computer portion could be put together for $500. Figure 9-3 shows signal-averaged luminescence decays acquired with their system. Peterson *et al.* (1979) have reported subnanosecond deconvoluted rise time data measured with the system. Suzuki *et al.* (1981) have reported a boxcar integrator which records multiple points per transient.

I. MICROCOMPUTERS

Micro- and minicomputers are currently revolutionizing instrumentation, and the area of excited state lifetime measurements is no exception. We feel that this area is so important and still so misunderstood by many scientists that we devote considerable space to it.

A wide variety of 8- and 16-bit microcomputer systems is available. The performance of these systems continues to rise as the price either remains relatively stable or decreases. Systems start at minimal single-board computers with random access memory (RAM), read only memory (ROM), a keyboard, a display, and an input–output (I/O) capability for a few hundred dollars; these are ideal for instrument control. Larger systems with high-level BASIC interpreters, video displays, full keyboards, and cassette-based programs and data storage facilities cost less than $1000, and a single floppy disk-based system can cost under $1500. A new generation of 16-bit microcomputers is becoming available at comparable cost. At the high end are 16-bit microcomputers such as the DEC LSI-11 and MINC systems with floppy disks, which cost $10,000–30,000. Many of these systems will largely or completely replace the use of central mainframe computers and increase worker convenience and productivity.

The software available even for low-cost systems is generally excellent, and some is even superior to that available from the large minicomputer manufacturers. Excellent BASIC interpreters are standard on many machines and come permanently stored in ROMs as the operating system. Several have superb video graphics including color. FORTRAN and BASIC compilers are available for a modest cost, as are excellent disk-operating systems such as CP/M. PASCAL compilers and interpreters, as well as the unique and very fast FORTH language, are available for a variety of systems. Thus, in many cases existing programs can be transferred more or less directly to mini- or microcomputers.

There are a variety of microcomputer central processing units (CPUs). The most popular 8-bit ones include the 6502, 6800, 8080, and Z80 single-chip CPUs. Functionally, the similar 6502 and 6800 are perhaps slightly faster than the 8080 and Z80 processors, although caution must be exercised since speed can really be defined only in the context of a specific problem. The 6502 and 6800 are probably harder for most users to program efficiently. In the 16-bit area there are the Intel 8086 series, the Zilog Z8000 series, the Motorola 68000, and the multichip Digital Equipment Company LSI-11 chip set. The newer 8086 and especially the Z8000 may be faster than the older LSI-11's. Thirty-two-bit processors are rapidly approaching the marketplace at this writing. Several of these processors are sold in computers that run at different clock frequencies. Thus, a 4-MHz Z80 system can generally execute all programs at twice the speed of a 2-MHz system running the same program.

In terms of program speed there is one caveat. When using commercial compilers or interpreters, never assume that software running on a more powerful processor will be faster. For example, the Z80 holds the 8080 processor's instruction set plus a whole range of more powerful instructions. Most Z80 BASIC interpreters execute at virtually the same speed (normalized to the clock frequency) as the 8080 ones because 8080 interpreters were moved directly to the Z80 (a trivial task) with little or no program optimization. Also, a currently popular 16-bit 8088-based microcomputer with a 5-MHz clock using the ROM-based Microsoft BASIC interpreters runs benchmark programs slower than a Microsoft BASIC interpreter with a 4-MHz 8-bit Z-80 system. Apparently the code was translated to the 16-bit processor with no optimization, and differences in CPU architecture caused the translated code to run slower on the more powerful CPU. Compilers that can run with more than one CPU (e.g., 8080 and Z80) very likely generate only the lower-level code of the simpler CPUs which will then execute no more rapidly on the more powerful CPU.

The newcomer should not be turned away by the very low cost of many "hobbyist" microcomputers. Many hobby systems are extremely well engineered, conservatively designed, and have service records comparable to anything on the market. Further, many of these microcomputer companies have much better records in terms of dealing with single users than do large manufacturers.

The S-100 hobbyist computer bus has been standardized by the IEEE. This bus is exceedingly popular, and enormous numbers of S-100-compatible interfaces and subsystems are available including 64K-byte memory boards, floppy and hard disk systems, and ADC and DAC boards; generally their prices are only a fraction of the cost of similar systems from minicomputer manufacturers. In addition, very powerful 16-bit microcomputer CPU boards using 68000, Z8000, or 8086 series processors are available that run on the S-100 bus.

One important area deserves special comment: software. No matter how powerful the processor and interface capabilities no system will perform without software. There are basically three types of software: (1) machine and assembly language, (2) interpreters, and (3) compilers.

Machine language is the directly executable binary instructions of the computer. It is usually entered in binary, octal, or hexadecimal through a front panel switch bank or a resident monitor program. Avoid using it if at all possible. Assembly language is a mnemonic form of machine language. For example, MOV A, B stands for copy or move the contents of register B into A. This is more comprehensible than the actual binary instruction: 01111000.

Programs written in these mnemonics are called assembly language programs. Assembly language programs are not in the computer's native language, and they must be translated into machine language by a program called an assembler. An assembler is generally a large program which may take up 4000–32,000 bytes of memory, but plunging memory costs makes this almost unimportant. Assembly language programs are relatively easy to document, modify, combine, or relocate. Any serious machine language programming should be done through an assembler. For the newcomer good assembly language programming requires a good understanding of the processor's native instruction set. Even with an assembler, machine language programming is slow, tedious, and error-prone; it is not to be undertaken lightly. The advantages of machine language are that programs written in it can be executed faster than any other type of program and they give the programmer exquisite control over the system; this makes them ideal for sophisticated interface applications and where speed is of the essence.

Interpreters and compilers are high-level languages that understand sophisticated statements like $X = (A + B) * C$ and translate them into the extremely complex set of machine language instructions necessary to carry out the operations. In an interpreter, the original source program (i.e., the one readable by humans) is stored at all times and is translated line by line into machine code only during execution. Since the source is always present, errors are easy to correct once they are encountered. Compilers are more complex programs which translate the entire original source program into a single block of machine-executable code that is run later. Compilers and interpreters each have advantages and disadvantages.

The disadvantages of compilers include the speed of program development and debugging. The compilation process is slow and virtually necessitates a disk operating system. Further, when an error occurs during the running of a compiled program, it is frequently unclear where the error arose; only the machine code is present, and this bears no clear relationship to the original source program. Thus, debugging is tedious, especially since to make even the tiniest change requires reloading the source, editing, and recompiling it. Once a running machine language program is generated, however, it can run typically three to ten times faster than the equivalent interpreted program. Further, it is easy to build up a library of routines with a compiler, which then greatly accelerates program development.

Interpreters are slower but much easier to debug since, generally, the compiler points out the line containing the error and it can be corrected on the spot. Interpreters do take up more memory at run time since the large interpreter and source programs must be coresident, but with the compiler only the more compact machine language module is required. This last disadvantage of interpreters is frequently illusory, however, since development of the compiled program generally takes as much or more memory as the interpreter itself.

To summarize, interpreters are definitely to be preferred for small-program development where execution time is not critical. Compilers are preferred if execution speed is important, the programs are complex, or a variety of preexisting library functions will be used. Currently, the most popular interpreters and compilers are BASIC and FORTRAN, respectively. Efficient BASIC compilers are now available. FORTH is an excellent semicompiled language. PASCAL is another excellent language, as it is highly structured and this promotes good programming. It is available in compiler and interpreter versions.

We turn now to some current applications of microcomputers in lifetime measurements. They can run from simple instrument controllers and data loggers all the way up to complete data reduction display systems rivaling the performance, for this application, of many mainframe computers.

Examples of computerized boxcar integrators are given in Section H. Also, all the simulations, deconvolutions, and plots prepared for this book were done on low cost microcomputers with a digital plotter.

We have described a microcomputer interfaced to a Tektronix ultra-high-speed transient digitizer (Turley, 1980). The digitizer records 512 points in a 5-nsec single shot. The data format is not suitable for direct processing, however, because of the obscure format. With a floppy disk-based 8-bit microcomputer, the data could be transferred to the computer, converted to a single-valued waveform, and reduced by any standard data reduction method. Excluding the training of the graduate student during this project, the total hardware costs were a fraction of the cost of the manufacturer's system.

Most manufacturers of SPC systems will supply dedicated and powerful 16-bit microcomputer systems for data reduction and graphics. Eight-bit micros are also useful for deconvolution. J. E. Löfroth (1980 personal communication) has interfaced a Commodore PET 8-bit microcomputer with the multichannel analyzer of his SPC system. The system, running BASIC, performs single- and double-exponential fits in ~3 and ~15 min, respectively. This compares with 30 sec and 2 min for a commercial 16-bit system. Fifteen minutes to acquire an SPC decay curve is not uncommon, and frequently much longer is required. Thus, the slow speed of the processor is not the limiting factor in data acquisition.

There are some basic rules of computer interfacing that should be studied by the newcomer before purchasing any system or trying to interface an instrument. Things have become much easier since the author purchased his first microcomputer in 1975, but the following points still remain relevant:

(1) Direct connections to the computer bus should be made only by those with extensive experience in digital electronics and with considerable insight into hardware–software interactions.

(2) Whenever possible, and certainly for the less knowledgeable, interfacing should be done through the computer's serial RS-232 ports, an IEEE-488 bus, or parallel ports. Many instruments such as MCAs are set up to communicate directly with computers through these ports, and no further interfacing is necessary. Also, a rapidly increasing number of instruments interface directly with the IEEE-488 bus. These include digital plotters, voltmeters, frequency counters, transient recorders, and oscilloscopes.

(3) Always have the manufacturers verify that instruments to be interconnected are compatible. For example, the popular MINC minicomputer will not accept signals from certain MCAs because of unacceptable control characters. In another example, an analog-to-digital converter

board that interfaces with "virtually all microcomputers" was added to the bus of a Radio Shack TRS-80 computer. It worked occasionally. After a week of work the problem was traced to an ~10–20-nsec timing skew in the bus signals. Reconfiguring the boards solved the problem, but the delay was costly.

(4) Never accept on faith a manufacturer's word that he will produce a given product. Delays in the mini- and microcomputer areas have been disastrous to many users who accepted a supplier's commitment. Most manufacturers are not dishonest, but development of new hardware and software is incredibly complex.

(5) Unless the application calls strictly for a limited function machine, always buy an upgradable system. There is an excellent chance that you will want to do something else that will require more memory or other additions.

(6) Always permit far more time for software development than you think it will take. This is especially true for machine or assembly language programming.

(7) Always write interface software in a high-level language unless it is very trivial or the speed of an assembly language program is essential. It is then easier to debug, to modify and update, and to be understood by others.

(8) Machine and assembly language programming is much harder to learn and write than many newcomers appreciate. It is great fun for many and provides exquisite control of what is going on, but allow considerable time if you have never undertaken such a project before and do not have expert advice handy.

(9) Always obtain all the software you need at the beginning. Minimum software is an interpreter. For machine language development, a good assembler and a debugging program are essential.

(10) If you have to do your own repairs, a modular system with socketed integrated circuits is invaluable. Many single-board computers cost less, but when any element fails, the board becomes nonfunctional. Trouble shooting is then a nightmare.

(11) Analyze closely the initial and long-term costs. At this time the update service of one minicomputer manufacturer for their operating systems and FORTRAN package is >$1000/year. The complete purchase price of a FORTRAN or PASCAL compiler for an 8-bit microcomputer is <$500. The minicomputer manufacturer's software is usually excellent, but is the extra performance worth the extra cost? As another example, the cost of repair of a minicomputer memory board was estimated at $400, even though the problem was only a $10 memory chip which the manufacturer would not supply. The cost of a 32K memory board for an S-100 microcomputer is <$360.

(12) If possible, go with a disk-based system. If you start out with a cassette-based system, you will almost certainly want to go to a disk system later. Generally better and more software is available for disk systems than for tapes. Possible exceptions are some of the very nice systems that have high-speed digital cassette recorders such as the Hewlett-Packard 85A.

(13) Ideally, obtain both a video terminal, with a good editor, and a printer. Very powerful video word processors are available for most microprocessor systems. These editors permit adding, deleting, or directly modifying text on the screen so that the changes can be seen. Editing on a video screen is far faster and less error-prone than on a printer. Indeed, the manuscript for this book was composed on one. However, debugging a long program exclusively on a video monitor where only a limited number of lines are available at one time can be exasperating. Hard copy printouts are invaluable for such problems. Also, you will always want some results printed.

(14) Never assume software will be upwardly compatible with the next generation of CPU. For example, when the Z80 microprocessor was first introduced, it was widely claimed that all 8080 software would run on it since the 8080 instruction set was a subset of the Z80. This was not always true. Execution times of the functionally same instructions were different, and one flag bit in the 8080 was used slightly differently than in the Z80.

(15) Finally, if one does not have ready access to an individual knowledgeable in the area, there are a number of fine courses available in microcomputer interfacing and programming. Such courses can be invaluable and can save countless hours and grief in bringing a system on line.

10

Special Error Sources

A. INTRODUCTION

In this chapter we describe a number of bothersome experimental problems. The problems of wavelength and positional timing effects in PMTs as well as a number of ways to circumvent these problems are discussed; these sections are extensive, as the problems are frequently the limiting factor in short-lifetime measurements. The serious, but frequently overlooked, puzzling problems of cable termination are demonstrated. The severe, sometimes even terminal, problem of electromagnetic interference is discussed. Self-absorption errors are described, and a method of compensation for long-term drift in the excitation profile is developed. We conclude the chapter with sections on triggering problems in general, pile-up problems in single-photon counting measurements, and the subtle but frequently overlooked problem of sample purity. We have already discussed in some detail (Chapter 7) the special problem of RC time constant effects.

B. WAVELENGTH EFFECTS IN PHOTOTUBES

Perhaps one of the most insidious and vexing problems associated with the deconvolution problem is the variation in PMT response with wavelength. The temporal response of many PMTs depends on the wavelength of the excitation light. This effect arises because the kinetic energy of the photoelectrons ejected from the photocathode is wavelength dependent. The maximum kinetic energy E_k of the ejected electron is given by $E_k = h\nu - W$, where W is the work function for the photocathode. Thus, an electron ejected by a red photon can start out with 1 eV less kinetic energy than an electron ejected by a near-UV photon striking the same photocathode. On the average, a photoelectron ejected by a UV photon starts with a higher initial velocity than one ejected by a red photon. This velocity difference causes the current pulse produced by the red photon to arrive later than the pulse from the blue photon. This effect is called the wavelength variation of the PMT transit time and manifests itself as a time shift seen at the phototube anode for otherwise identical light pulses of different wavelengths. Although this energy difference may seem small, it can cause a time shift of a nanosecond or more.

The situation is even more complicated, however. The distribution of the velocities of the photoelectrons may differ for different excitation wavelengths. Thus, not only can there be a wavelength dependence on the transit time, but also the pulse shape can vary with wavelength. For example, Wahl *et al.* (1974) found shifts of ~1.3 nsec for 317-versus 568-nm excitation of a 56DUVP03 PMT, while the pulse width decreased from 2.85 to 1.58 nsec. Thus, with increasing excitation wavelength, their pulse was progressively shifted to a longer time but became narrower. It still remains unclear whether this is a real effect or whether the tunable light source used exhibited temporal variations with wavelength. Other workers have, however, observed shifts varying from 50 psec to ~1 nsec. We now describe a variety of correction schemes that have been used and discuss the merits of each.

1. Trial-and-Error Time Shifting

Conceptually the most obvious approach is to measure the excitation $E(t)$ and decay $D(t)$ profiles each at their respective wavelength. Then one deconvolutes the decay parameters from the observed $E(t)$ and $D(t)$ for a range of time shifts of one curve relative to the other. The shift yielding the smallest χ^2 is taken as the necessary shift, and the corresponding decay parameters are the best fit.

This approach works well if the PMT characteristics can be approximated accurately by a transit time shift but fail if the PMT output band shape varies with excitation wavelength. Computationally, this shifting technique is generally too time-consuming to use with every decay curve, especially if the decay kinetics are complex. It is, thus, common to calibrate an instrument at a number of different wavelengths and merely read the necessary shift for a specific experiment from this curve and carry out the necessary shift before deconvolution.

2. Quantum Counters

A luminescent quantum counter (QC) is an optically dense fluorescent screen, usually a solution of a highly luminescent dye, placed before an optical detector such as a PMT. If the photon yield and emission spectrum of the dye are wavelength independent, then the response of the detector will become independent of the excitation wavelength (Parker, 1968; Demas and Crosby, 1971; Taylor and Demas, 1979; Mandal et al., 1980). In the present context, if the dye has a decay time that is independent of the excitation wavelength, then regardless of the excitation wavelength the PMT will always see the same distribution of light, and any wavelength effects on the transit time will be avoided.

Thus, the QC experiment is as follows: A counter material is selected that absorbs both the sample emission and the excitation. Instead of viewing the flash and decay directly, the PMT now views them both after shifting by the quantum counter solution to obtain $E(t)$ and $D(t)$. These signals are delayed in time and broadened by the lifetime of the quantum counter in the same manner. By sending the flash and decay pulses through the QC we have introduced an identical linear distortion of both signals. But, as shown in Section 7.B, the same linear distortion applied to both flash and decay data has no affect on subsequent deconvolution.

An excellent application of the use of a QC was reported by Lewis et al. (1973) who used a POPOP QC. They were able to show that the major wavelength shift seen in their single-photon counting nanosecond fluorimeter was a wavelength-dependent transit time.

The major disadvantage of this approach is its enormous loss of sensitivity in the measurement of $D(t)$. The already weak sample emission is further attenuated by the quantum yield of the QC and the small solid angle of QC emission intercepted by the PMT. The number of reliable QCs is limited, and it is hard to guarantee the restriction that the QC's decay curves must be independent of the excitation wavelength. Radiation trapping and, thus, decay times of strongly self-absorbing dye solutions can be highly dependent on the sample optical density and penetration of the excitation beam into the sample. Further, such widely accepted counters as rhodamine B have efficiencies which vary with wavelength (Taylor and Demas, 1979) and, thus, there may be problems with the wavelength dependence of their decay curves. Another problem is that the QC decay time must be short to minimize the distortion of the curves and the resulting loss of temporal information.

The major advantage of the QC approach is, of course, its complete elimination of the wavelength effect in the PMT. Its great loss of sensitivity, however, largely eliminates its use in the measurement of both $E(t)$ and $D(t)$.

A frequently overlooked use of the QC, however, is in direct determination of the transit time shift at different wavelengths. One needs merely to measure a decay curve of the flash at different excitation wavelengths both with and without the QC in place. If the decay curve shape is independent of the excitation wavelength with the QC present, then the flash profile is independent of the excitation wavelength. Then, variations in the time shifts between the two curves versus the wavelength are a direct measure of the shift versus the wavelength. These shifts can then be used with a table listing in subsequent deconvolutions. The author is unaware of such a use of QCs, but the approach appears to be a valuable one.

3. Use of Excitation Flash

Perhaps the most common approach to measuring shifts is to use the excitation source itself as the reference. With continuum sources (H_2 and D_2 arcs) the assumption is made that the true temporal profile of the flash is the same at all wavelengths. Thus, to determine the flash profile without any time shift, one merely tunes the excitation source to the sample emission wavelength before measuring $E(t)$. Then, if the excitation shape is wavelength independent, this $E(t)$ has the same pulse shape as if it had been measured at the true excitation wavelength but was being viewed at the emission wavelength. In this way transit time shifts with wavelength are eliminated.

The major disadvantage of this approach is the difficulty of being sure that the flash profile is wavelength independent. It appears to be true that a D_2 flash indeed satisfies this criterion over a wide wavelength range, but only if the gas is kept scrupulously clean. The presence of traces of N_2 can produce a spurious stretching of the pulse, especially in the 300–400-nm region. Finally, the method cannot be used with noncontinuum sources such as lasers. Even tunable lasers have pulse shapes and delays that can vary strongly with wavelength.

4. Derivative Methods

If a reference compound with a known lifetime is available, it is possible to calculate from its observed decay curve and decay time what the flash would have looked like had it occurred at the monitored emission wavelength. The necessary expression is

$$E(t) = R(t) + \tau_R[dR(t)/dt] \tag{10-1}$$

where $E(t)$ is the flash profile, τ_R is the decay time of the reference, and $R(t)$ is the reference decay curve. This expression is derived directly from differentiation of Eq. (4-3), where $D(t)$ is now $R(t)$. Wahl *et al.* (1974) have suggested the following formula for evenly spaced data:

$$E(t_i) = R(t_i) + (\tau_R/2h)[R(t_{i+1}) - R(t_{i-1})] \tag{10-2}$$

where h is the time interval between points. This expression is derived from Eq. (10-1) by fitting a parabola through the three points $R(t_{i-1})$, $R(t_i)$, and $R(t_{i+1})$ and evaluating analytically the slope of the resultant parabola at t_i.

We demonstrate the use of the derivative method by some simulated examples. Figure 10-1 shows the extraction of $E(t)$ from a series of simulated single-photon counting decay curves. With $\exp(-t/5.5) - \exp(-t/6.6)$ as $E(t)$, an exact $R(t)$ was calculated for $\tau_R = 10$ nsec. $R(t)$ was scaled to 10^N counts ($N = 2, 3, 4, 5, 6$) in the peak channel, and Poisson noise was added. With the use of Eq. (10-2), $E(t)$ was calculated. For 10^3, 10^4, 10^5, and 10^6 peak counts in $R(t)$ the resultant back-calculated $E(t)$'s are shown in Fig. 10-1. Except for the small amount of noise, the $E(t)$ calculated from the $R(t)$ curve with 10^6 peak counts agrees essentially perfectly with the original noise-free $E(t)$. Thus, for high enough quality data, successful reconstruction of $E(t)$ is possible. It is clear, however, that the recon-

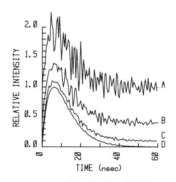

Fig. 10-1. Simulated reconstructed $E(t)$'s using a reference emitter and Eq. (10-2). The reference has $\tau_R = 10$ nsec. The peak counts in the reference decay curve $R(t)$ before the addition of Poisson noise are as follows: $R_{peak} = 1000$ (A); 10,000 (B); 100,000 (C); and 1,000,000 (D). Curve A is displaced by 1.0, curve B by 0.4, and curve C by 0.1 units to improve viewing. There is no significance to the relative amplitudes of the curves. Curve D agrees essentially perfectly with the assumed $E(t)$ profile.

structed $E(t)$'s become progressively less satisfactory as the peak counts in $R(t)$ decrease. Indeed, even for a peak $R(t)$ of 10^4, which is routine in single-photon counting data, the reconstructed $E(t)$ is quite noisy. For a peak in $R(t)$ of 1000, the reconstructed $E(t)$ can be considered little better than a disaster. For a peak in $R(t)$ of 100, the reconstructed $E(t)$ was essentially all noise with only a suggestion of the correct underlying waveform; it is not even reproduced in Fig. 10-1, since it is so noisy that it would span all the other curves shown.

The exceptionally noisy character of the reconstructed $E(t)$ is not, however, the full story. The real question concerns the accuracy of the resultant deconvolutions using these reconstructed $E(t)$'s. Intuition suggests that the results will be very poor unless very large counts ($\geq 10^5$) are used for the peak of $R(t)$. Surprisingly, at least if moments method 1 and to a lesser extent the phase plane method are used to deconvolute the decay times, the calculated lifetimes will be in remarkably good agreement with the correct ones. Table 10-1 shows the results of simulated deconvolutions using different peak values in the $R(t)$ and $E(t)$ curves (R_{peak} and E_{peak}, respectively) before the addition of noise. For these calculations the same exciting flash $E(t)$ of Fig. 10-1 was used with $\tau_R = 10$ nsec. Sixteen simulations were carried out for each set of conditions, and the mean lifetime and standard deviation are reported. Method A involves

TABLE 10-1

Results of Deconvolution by Means of a Reference Emitter

D_{peak}	Method[b]	Calculated Mean and Standard Deviation		
	$R_{peak}{}^a$	100	1000	10,000
100	A	4.921 ± 0.248	5.006 ± 0.204	4.933 ± 0.129
	B	5.286 ± 0.504	5.243 ± 0.372	5.015 ± 0.344
	C	5.274 ± 0.291	5.096 ± 0.267	4.944 ± 0.201
1,000	A	4.973 ± 0.216	4.981 ± 0.057	4.981 ± 0.050
	B	5.459 ± 0.667	5.008 ± 0.126	5.010 ± 0.099
	C	5.254 ± 0.320	5.002 ± 0.107	5.045 ± 0.068
10,000	A	4.951 ± 0.270	4.594 ± 0.059	4.985 ± 0.016
	B	5.462 ± 0.494	4.993 ± 0.124	5.012 ± 0.068
	C	5.306 ± 0.198	5.021 ± 0.105	4.986 ± 0.022

[a] D_{peak} and R_{peak} are the peak channel counts in the decay and the reference, respectively. The sample lifetime used to simulate the decays is 5.000.

[b] Methods A and B use Eq. (10-2) to calculate $E(t)$. In method A, moments method 1 is then used to deconvolute the lifetime, and in method B the phase plane method of deconvolution is used. Method C used the integral phase plane method [Eq. (10-8)].

data reduction by moments method 1. The τ's calculated using $R(t)$ and $D(t)$ agree remarkably well with the expected lifetimes. Even for the apparently unusable $R_{peak} = 10^2$ or 10^3, the calculated τ's are quite acceptable. Method B is the same as method A, except that deconvolution was by the phase plane method.

Thus, if moments method 1 is used for deconvolution, the derivative approach of Eq. (10-2) yields excellent estimates of τ's free of any PMT time shift problems even for rather noisy $R(t)$ curves. It is clear, however, that using other deconvolution methods will not necessarily lead to such good results. Inspection of Table 10-1 shows that data reduction by the phase plane method consistently yields less precision than moments method 1, although for reasonably noise-free data ($R_{peak} = D_{peak} \geq 1000$) quite acceptable results are obtained. Except for very large R_{peak}, however, the $E(t)$'s back-calculated by the derivative method are very noisy and supply little direct information about $E(t)$. Further, there is no guarantee that other deconvolution methods will be inherently as insensitive to the derivative noise on the $E(t)$'s as the moments and phase plane methods.

Further, data taken by other than single-photon counting methods may not be amenable to the derivative approach. We have, for example, attempted to use the derivative approach for noisy data taken with a sampling oscilloscope system with weak emitters. We found that, after evaluating $E(t)$ from the reference and then attempting to deconvolute by a least squares method, the results exhibited great sensitivity to noise and the resultant lifetime calculations were unacceptable. Extensive signal smoothing of the $R(t)$ before evaluation of the derivatives led to improved but still unacceptable $E(t)$'s and deconvolutions. In light of the above results, it is unclear whether these results are a failure of the derivative method to handle the noise from the sampling system or a failure of the least squares method to handle the noisy derivative data. Thus, we do not recommend use of the derivative approach except for very high-quality data or for data reduced by moments methods.

Another problem with this derivative approach has been pointed out by Rayner *et al.* (1976), although their comments apply to any method using a reference emitter of known lifetimes. If erroneous lifetimes are used for the reference compound, quite respectable appearing deconvolutions (as judged by the residual plots and the chi-squares) can be carried out with the back-calculated $E(t)$, but the calculated decay times can be greatly in error. For example, using PPO in cyclohexane as a reference, they determined the decay time of POPOP in ethanol. For the same data they assumed a range of τ_R for the PPO reference, back-calculated the $E(t)$, and deconvoluted the data to obtain τ's for the POPOP. For $\tau_R =$

1.08 nsec they calculated $\tau = 1.18$ nsec, for $\tau_R = 1.28$ nsec they obtained $\tau = 1.38$ nsec, and for $\tau_R = 2.00$ nsec, $\tau = 2.13$ nsec. The weighted residual plots showed the deviations between the observed and calculated curves to be fit equally well in all cases. These results must not be treated as a general condemnation of the derivative or reference approach. Their warning applies only if $\tau_R \sim \tau$. For example, if $\tau_R = \tau$, whatever lifetime is assumed for τ_R before calculation of $E(t)$ will always yield τ_R, and the fit will always look normal. In the Rayner et al. (1976) example, POPOP and PPO have nearly the same lifetimes. Thus, the calculated τ for the unknown will closely track the assumed τ_R (as is observed) and the residual plots will be nearly the same for any assumed τ_R. If τ_R and τ are not too similar, however, the residual plots will begin to reveal the presence of errors in τ_R. This result stresses the care that must be exercised in the selection of standards. Further, more than one standard with different lifetimes should be used to check for possible errors.

5. Laplace Transforms and Moments Methods

The Laplace transform treatment can be modified to extract a time shift from the observed $E(t)$ and $D(t)$ if a simple time shift is the only distortion. The reader is referred to the original paper for details (Gafni et al., 1975). This method is useful for evaluating the time shift which can then be used in subsequent deconvolutions. It requires that the PMT exhibit only a simple time shift.

The moments methods also provide a variety of approaches to either eliminate or evaluate time shift errors. The moments method 2, which uses the radius of gyration, is completely insensitive to time shift. Moments indexing of orders greater than zero also inherently avoids time shift errors. These claims apply only if the wavelength-dependent distortion is a simple time shift between $E(t)$ and $D(t)$.

If the sample is characterized by a single-exponential decay, moments methods can be used to evaluate explicitly the time shift Δt. Moments method 1 depends on τ and Δt as follows:

$$\tau + \Delta t = (\mu_1/\mu_0) - (m_1/m_0) \tag{10-3}$$

where μ and m are the appropriate moments of the decay and excitation and Δt is the time shift. See Section 8.E for details of the moments methods. Thus, if τ is known from either a separate determination or from MM2, which is time shift independent, one can readily calculate Δt and apply it to all subsequent data before deconvolution.

6. Integral Phase Plane Method

The phase plane method can be suitably modified to handle the time shift problem. The PP equation is

$$W(t) = -(1/\tau)Z(t) + K \qquad (10\text{-}4a)$$

$$W(t) = D(t)\Big/\int_0^t E(t)\,dt \qquad (10\text{-}4b)$$

$$Z(t) = \int_0^t D(t)\,dt\Big/\int_0^t E(t)\,dt \qquad (10\text{-}4c)$$

One notes that only the running integral of $E(t)$ rather than $E(t)$ is actually required. If one monitors the decay curve $R(t)$ of a reference emitter, then the integral of $E(t)$ is given by

$$I(t) = \int_0^t E(t)\,dt = R(t) + (1/\tau_R)\int_0^t R(t)\,dt \qquad (10\text{-}5)$$

The significant feature is that, as $\tau_R \to \infty$, the expression simplifies to

$$I(t) = R(t), \qquad \tau_R \gg \text{flash width} \qquad (10\text{-}6)$$

Thus, merely by choosing a luminescent reference with a very long lifetime $R(t)$ is just the desired integral of the $E(t)$. In this case the lifetime of this reference does not even enter into the calculations. Thus, all one needs is a long-lived reference compound that emits at the same wavelength as the unknown. Under these conditions the PP equation becomes

$$D(t)/R(t) = -(1/\tau)\Big[\int_0^t D(t)\,dt/R(t)\Big] + K \qquad (10\text{-}7)$$

where $\tau_R \gg$ pulse width. Even if τ_R is not long enough to justify rigorously the approximation of Eq. (10-7), for long τ_R, the error associated with the $1/\tau_R$ term in Eq. (10-5) is small and the necessary correction can be made with great accuracy. The general PP equation is then Eq. (10-4a), but $W(t)$ and $Z(t)$ are given by

$$W(t) = D(t)/I(t) \qquad (10\text{-}8a)$$

$$Z(t) = \int_0^t D(t)\,dt/I(t) \qquad (10\text{-}8b)$$

where $I(t)$ is calculated from Eq. (10-5). This expression is correct for any τ_R. There is a benefit of the use of a long τ_R. If $\tau_R \gg$ pulse width, then $R(t)$ is just an exponential decay with a lifetime of τ_R. Thus, τ_R, even if originally unknown, can readily be evaluated from the decaying edge of the $R(t)$ curve by the standard semilogarithmic plot of intensity versus time.

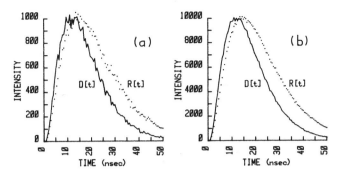

Fig. 10-2. Typical data used for extracting a lifetime using a reference emitter. In both sets of data, $\tau = 5$ nsec and $\tau_R = 10$ nsec. $R(t)$ and $D(t)$ are the reference and sample decay curves, respectively. In both cases Poisson noise was added. In (a), the peak counts on $D(t)$ and $R(t)$ were 1000, and in (b) the peak counts were 10,000, both before the addition of noise.

As an example of this approach, we apply the integral phase plane method to the single-photon counting data of Fig. 10-2. The experimental conditions are the same as for Fig. 10-1, and the sample decay curve has a τ of 5 nsec. Figure 10-3 is the integral PP plot calculated from these data. For the noisy data of Fig. 10-3a the error in τ is -4.8%, and for the more typical data of Fig. 10-3b the error is only $+0.2\%$.

Table 10-1 summarizes the results of lifetimes calculated using the integral phase plane method (method C). The results are in quite acceptable agreement with the correct values. For the conditions of Table 10-1, the PP method is not as good as the derivative method coupled with moments method 1 for deconvolution. The PP method consistently yields higher standard deviations, although the PP method yields quite accept-

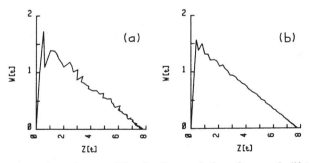

Fig. 10-3. Phase plane plots used for the deconvolution of a sample lifetime using a reference emitter and the integral method. The curves are generated from the data of Fig. 10-2. Using unweighted linear least squares fitting over the 12–75-nsec range, the errors in τ were -4.8% in (a) and $+0.2\%$ in (b).

able results for R_{peak} and $D_{peak} \gg 1000$, which are the conditions normally encountered. Several points should be made in comparing derivative and integral approaches. The integral approach is clearly superior to the derivative method coupled with phase plane data reduction, but it is somewhat inferior to the derivative method coupled with data reduction using moments method 1. Unlike the derivative approach, however, the PP plots are not very sensitive to noisy signals and can be used as good visual indicators of the success of the assumption of a single-exponential sample impulse response. Further, in this example the time domain for data acquisition was long enough virtually to eliminate cutoff errors in the moments calculation. The integral method, however, does not require cutoff corrections and, thus, is more convenient if a shorter time window is used so as to obtain better temporal resolution.

The question then arises as to what reference molecules to use. A variety of molecules given in Berlman's (1971) book should be suitable. In particular, anthracene, although short-lived, has a well-known lifetime, and pyrene has a long lifetime. For the orange near-IR regions ruthenium(II) complexes appear excellent. $[Ru(bpy)_3]^{2+}$ (bpy = 2,2'-bipyridine) is commercially available and has a 600-nsec lifetime in deoxygenated water. Its emission spans approximately 560 nm to >800 nm. At least at $\lambda \geq 600$ nm, the emission is characterized by a single-exponential decay. A similar approach with a $[Ru(bpy)_3]^{2+}$ reference was used by Peterson et al. (1979) to measure the luminescence rise times of a series of Rh(III) metal complexes. In spite of an ~1-nsec wavelength-dependent shift in their PMT and a 10-nsec-wide laser pulse, they were able to show that some of the rise times were ≤ 0.1 nsec, which agrees with data measured using very expensive picosecond methods (Kobayashi and Ohashi, 1982).

7. Broad-Band-Emitting Reference Compound

Rayner et al. (1976, 1977) have described a clever but somewhat limited approach. They selected a reference compound that emits at both the excitation and emission wavelengths. The method is cumbersome, involves extensive computations, and is not likely to be of broad usefulness. We describe its development here in part because it demonstrates the enormous power of the convolution integral approach.

It begins as a standard method in that one measures the excitation profile $E_{ex}(t)$ at the excitation wavelength by scattering part of the emission into the detector. The sample decay $D(t)$ is then measured at its emission wavelength λ_{em} while exciting at λ_{ex}. If there were no PMT wavelength effect, one could then directly deconvolute the results from $E_{ex}(t)$ and $D(t)$. To eliminate the PMT wavelength effect, one then puts in a reference molecule that emits at both λ_{em} and λ_{ex} for the unknown. The reference is excited at shorter wavelengths than λ_{ex}, and the decay curves for the reference are measured at λ_{ex} and λ_{em} to yield $R_{ex}(t)$ and $R_{em}(t)$, respectively. The differences between $R_{ex}(t)$ and $R_{em}(t)$ now carry the necessary information concerning the wavelength dependence of the PMT to correct the $E_{ex}(t)$ and $D(t)$ data.

The derivation proceeds as follows: In terms of convolution integrals (Chapter 6), for the unknown sample we have

$$D(t) = H_{em} * e_x * i_x \tag{10-9}$$

$$E_{ex}(t) = H_{ex} * e_x \tag{10-10}$$

where $e_x(t)$ is the true flash profile used to excite the unknown sample at λ_{ex}, H_{em} and H_{ex} are the PMT detector system's impulse response when excited at λ_{em} and λ_{ex}, respectively, and i_x is the emission impulse response of the unknown sample. Note that H_{ex} and H_{em} do not have to be the same.

For the reference sample we have

$$R_{em}(t) = H_{em} * e_R * i_R \tag{10-11}$$

$$R_{ex}(t) = H_{ex} * e_R * i_R \tag{10-12}$$

where i_R is the luminescence impulse response of the reference and e_R is the flash profile used to excite the reference. It is assumed explicitly that i_R is the same at λ_{ex} and λ_{em}. Failure of this assumption totally invalidates this procedure. It is not necessary for $e_{ex}(t)$ to equal $e_R(t)$.

We can now use Eqs. (10-9) and (10-12) to evaluate $D * R_{ex}$ as follows:

$$D * R_{ex} = (H_{em} * e_x * i_x) * (H_{ex} * e_R * i_R) \tag{10-13}$$

Since the convolution integrals satisfy the commutative and associative properties, Eq. (10-13) can be rearranged to yield

$$D * R_{ex} = i_x * (H_{em} * e_R * i_R) * (e_x * H_{ex}) \tag{10-14}$$

which, from Eqs. (10-10) and (10-11) becomes

$$D * R_{ex} = i_x * (R_{em} * E_{ex}) \tag{10-15}$$

The functional form of Eq. (10-15) is identical to Eq. (6-13) which is the normal convolution relationship between a flash and decay. Now, however, $D * R_{ex}$ replaces the normal decay and $R_{em} * E_{ex}$ replaces the flash. Thus, if we think of $D * R_{ex}$ as a decay and $R_{em} * E_{ex}$ as a flash, we can use any standard deconvolution procedure to extract the desired i_x. This result is even more remarkable since we do not need to know e_R, e_x, or τ_R. Further, e_R and e_x do not even need to be the same. This last point is very important from a practical standpoint, since many sources have temporal behavior that varies strongly with wavelength.

The use of Eq. (10-15) is not as simple as indicated above. Rayner *et al.* state that the errors in the derived quantities $D * R_{ex}$ and $R_{em} * E_{ex}$ are no longer characterized by single-photon statistics and are, thus, not amenable to normal nonlinear least squares fitting. Rayner *et al.* (1976, 1977) recommended using Fourier transforms to deconvolute the data; however, since these authors elsewhere recommended against the use of deconvolution by Fourier transforms, it is unclear what the best procedure is.

Fortunately, given its complexity, use of the method of Rayner *et al.* (1976, 1977) can usually be avoided. Most simple systems can be adequately treated by trial-and-error fitting of a simple time shift of the $E(t)$, derivative, integral, or moments method, which can eliminate the effect. Only where the greatest accuracy is required with complex decay kinetics (e.g., sums of exponentials) and the PMT exhibits more complex effects than just a time shift does this approach appear attractive.

8. PMTs without Time Shift

The ideal way to carry out such measurements is to use PMTs without a time shift. It has been noted that some tubes are worse than others and that the delays are critically dependent on adjustments of the focusing and dynode voltages. It now appears, however, that the lowly and much maligned side-view squirrel cage PMT may be virtually free of such effects (Andre *et al.*, 1979). The principal disadvantage of such tubes is a lower sensitivity than that found in many end-on tubes. However, even with single-photon counting instruments, improvements in low-level discriminators now permit side-view tubes to be used without preamplifiers. The original study was made with a PMT especially designed for single-photon counting, but it seems plausible that many other tubes of this type function similarly. Kinoshita and Kushida (1982) have shown that the R928 side-view tube has a transit time shift that varies by only 19 psec/100 nm over the 154–620–nm range. Corrections for the count rate are, however, required. Finally, the relatively new microchannel plate PMTs have extremely short internal distances. It seems likely that they may be virtually free of wavelength effects, although the author is unaware of any studies that have been made on this characteristic.

C. PHOTOMULTIPLIER POSITIONAL EFFECTS

A particularly insidious effect has been observed in the use of PMTs. The transit time through the PMT is highly dependent on the portion of the photosensitive surface that is illuminated, and for both end-on and side-view PMTs the differences in transit times through the tube can be a nanosecond or larger (Kinoshita *et al.,* 1981; Kinoshita and Kushida, 1982). Thus, even if the PMT has no wavelength dependence on its transit time, very large errors can creep into subsequent deconvolutions if $E(t)$ and $D(t)$ are measured under conditions where different portions of the photosensitive surface are illuminated.

For successful deconvolutions at short times it is essential that $E(t)$ and $D(t)$ be measured using the same portion of the photocathode. This is routinely arranged by placing the reflector or scatterer for the $E(t)$ measurement in such a way as to mimic the sample emission geometry. For example, in an optically dense front-surfaced emission, the scatterer replaces the front of the sample cell. For an optically dilute sample in a square cell viewed at right angles a mirror or, better still, a diffuse reflector (e.g., ground glass, MgO, or $BaSO_4$) is set at a 45° angle at the center of the cell. A scattering solution [e.g., Ludox (colloidal silica) or glycogen] of the same optical density as the sample can be used, but caution is advised as these can emit.

The use of a mathematical deconvolution that eliminates time shifts is not recommended for avoiding positional effect errors. Depending on photocathode illumination, $E(t)$ and $D(t)$ could be made up of the superposition of a number of different time-shifted pulses. If the illuminations are the same, there will be no problem since $E(t)$ and $D(t)$ are broadened out in an identical manner. If either one is different, however, the superpositions will be different for the two, and the observed curves will be inappropriate for deconvolution with each other.

It should be noted that one advantage of the use of quantum counters in front of the PMT is that the diffuse QC emission uniformly illuminates the photocathode. This uniform illumination holds for both the $E(t)$ and $D(t)$ measurements. Thus, the QC smoothes out variations in the positional dependence of the PMT transit time. This smoothing can, in principle, virtually eliminate these effects. As pointed out in Section B.2, however, reliable QCs are difficult to come by and their use extracts a severe penalty in system sensitivity. If and when truly reliable quantum counters can be obtained, they can be used very profitably to reduce positional effects.

Another interesting positional effect has been observed by Hartig and

Saver (1976). These authors found that, when using an end-on tube with a semitransparent photocathode, the pulse shape differed depending on whether the center or edge of the photocathode was excited. When the center was illuminated, a weak extra peak appeared ~7 nsec before the onset of the main lamp peak. This pulse appeared to arise from photons that passed through the photocathode and struck the first dynodes causing photoelectron ejection. Since these photoelectrons did not have to traverse the large distance between the photocathode and the first dynode, this pulse arrived at the anode before the main photocathode-derived signal pulse. This ancillary pulse could be virtually eliminated by illuminating the photocathode so that no transmitted light directly struck the dynodes. This ancillary peak contained ~0.35% of the total counts and, if $E(t)$ and $D(t)$ were determined under greatly different conditions, differing contributions from this component could certainly cause deconvolution inaccuracies. Tubes with opaque photocathodes should be largely immune from this effect.

D. SAMPLE PHYSICAL PLACEMENT

Even ignoring variations in transit time associated with different positions on the PMT surface, the speed of light causes very short time resolution effects. Since the speed of light is 3×10^{10} cm/sec, an optical difference of 3 cm between the path traversed to the PMT of $E(t)$ versus $D(t)$ corresponds to 100 psec.

Even the sample refractive index can be important. Thus, the speed of light depends on the refractive index n of the media through which it passes and is given as c/n. If the sample emission passes through 0.5 cm of solution with $n = 1.5$ versus 0.5 cm of air, there will be ~10 psec of time shift difference between the two waveforms. Such a situation might result when $E(t)$ is reflected off a mirror at what would have been the cell center for the measurement of $D(t)$. Thus, for the very shortest time resolution, effective optical path lengths must be controlled very closely or a data reduction method that is insensitive to time shifts between $E(t)$ and $D(t)$ must be used.

E. CABLES AND TERMINATIONS

For workers used to working at low speeds (<10 MHz), the transition to studies in the low-nanosecond domain can be most traumatic. No longer can a piece of coaxial cable be treated as a simple connection between two circuits and a resistor as a simple device for converting the phototube current to a voltage. Simple *RC* time constant effects (Chapter 7) can be completely overshadowed by the far more damaging effect of improper cable termination.

For high-speed phenomena a cable or even a wire represents a transmission line with a certain characteristic impedance (AC resistance). Typical cable impedances are 50, 75, and 91 Ω. When a fast signal is sent down such a transmission line, it propagates smoothly along the cable at a speed of ~0.5 times the speed of light—as long as the cable is uniform. When the signal arrives at the far end, if it finds a resistor to ground that has the same impedance as the cable, the signal considers the resistor a continuation of the cable and is absorbed by it. A device monitoring the signal across the resistor will then see a faithful reproduction of the pulse. The cable in this case is considered to be terminated at its characteristic impedance.

If, however, the termination resistor is either appreciably smaller or larger than the characteristic cable impedance, there will be serious problems. The cable is then improperly terminated, and some of the signal is reflected back down the cable either in phase or out of phase with the input signal. This reflected signal can then bounce back and forth down the cable several times before the oscillations are damped out. For a step excitation, for example, a severe overshoot and damped ringing can occur on the observed waveform or the waveform can monotonically approach the steady state values in a series of steps. The duration of each step and oscillation is equal to the round-trip time of the signal in the cable.

These effects are always present, but they are not noticed at low speed for two reasons. First, the time required for the oscillations to die out is usually short compared to the monitoring times. For example, even with a 1-m cable, the round-trip time is ~10 nsec so that, even if ten oscillations are required to reach a steady state, the signal is stable in only 100 nsec. The second reason the effects are not noticed is that for slowly rising input signals the observed waveform will be a composite of many overlapping reflections. Thus, although these reflections will somewhat slow or accelerate the rise and fall of the observed slow signals, the effect will be smooth and rarely noticed. Only if the round-trip time in the cable is comparable to, or longer than, the rise or fall time of the signal do they become noticeable.

Such perturbations of waveforms are extremely useful in an area called time domain reflectometry (TDR). A fast-rising pulse is sent into one end of a cable, and the reflected waveform is monitored at the injection point. Perturbations on the reflected signal can be used to judge the quality of the connections in the cable and of the cable itself. Further, when there is a physical break or short in the cable, a characteristic reflected signal results and the time required for the reflection to return can be correlated with the distance to the break. TDR is routinely used to locate breaks in underground cables, so that workers need only dig up the damaged section. In luminescence, however, these reflections are a real nuisance.

To understand one of the reasons why proper termination is so bothersome, we point out that the most common cable impedance is 50 Ω. Occasionally, 75- and 91-Ω impedances are seen. These are very low impedances and represent the highest resistance that can be used on a properly terminated cable. But the PMT current is converted to a voltage across this load resistor. Thus, these small characteristic impedances greatly limit the system's sensitivity. For example, to obtain a 100-mV signal across a 50-Ω load resistor requires 2 mA. Such a large current can be supplied only transiently or not at all by the PMTs.

We now present some examples of the adverse effects of improper cable terminations on both fast and relatively slow waveforms. The example concerns the use of an ultrahigh-speed transient digitizer (Tektronix 7912) which has a 50-Ω input impedance for use with 50-Ω cable. It is, however, expensive and fragile if abused. Therefore, we frequently use a relatively slow 1-MΩ input resistance scope on the cable to permit independent monitoring of the signal waveforms. Figure 10-4 shows several possible configurations for connecting the transient digitizer and the monitor's oscilloscope (A–C). In all these cases a suitable 50-Ω termination is on the cable. Configuration D in Figure 10-4 shows a system in which the extra cabling and monitor's scope are omitted, but the 50-Ω termination is replaced with a 30-Ω one. In all cases, except for the small loss of signal in D in Fig. 10-4, a dc circuit analysis would indicate no differences among the four configurations. Under pulsed conditions, however, they are quite different.

In Fig. 10-5 curves A–C show the response of the configurations A–C, respectively, in Fig. 10-4 to the laser-excited decay of POPOP ($\tau \sim 1.2$ nsec). In Fig. 10-5 curve A shows the clean, faithful reproduction possible when the signal arrives at the end of the cable and a proper termination simultaneously. In Fig. 10-5 curves B and C show severe reflections that result if the cable is improperly terminated at the end. Note that the first reflected spike is delayed twice as long when the monitor scope is 12 ft rather than 6 ft beyond the transient digitizer. Further, note that, for the improperly terminated cases, significant undershoot (i.e., negative-going signals) results. The signals are severely distorted for >200 nsec after the actual end of the pulse. Especially with the 12-ft cable, the cable ringing

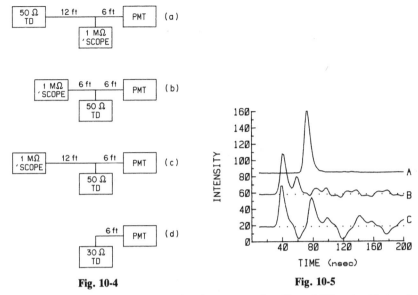

Fig. 10-4   **Fig. 10-5**

Fig. 10-4. (a)–(d) Configurations used to demonstrate the effect of different cable termination schemes with fast waveforms. PMT, Photomultiplier with no load resistor; TD, 50-Ω input resistance high-speed transient digitizer. 'SCOPE, 1-MΩ input resistance monitor oscilloscope. The lengths of the 50-Ω characteristic impedance cables are indicated in feet. The length of the "T" connection to the middle device is negligible. In (d) the input impedance of the digitizer was reduced to 30-Ω by a parallel terminator.

Fig. 10-5. Response of the different configurations of Fig. 10-4 to the laser-excited emission of POPOP. Curve A corresponds to Fig. 10-4a, B corresponds to Fig. 10-4b, and C corresponds to Fig. 10-4c.

pulses are nearly as large as the primary pulse, which would wreak havoc with any type of pulse detection equipment. Finally, the overall complexity of the waveform is caused by the fact that the cable is not terminated at the phototube. Thus, reflections once started bounce back and forth between the monitor and the PMT.

In Fig. 10-6, curves A–C show the response of configurations A–D in Fig. 10-4 to a laser-excited 400-nsec decay. The properly terminated curve A has a clean rise and decay, but the improperly terminated curve B shows the severe ringing and slow rise time caused by the unterminated 12 ft of cable to the monitor scope. In this case ringing is still visible 500 nsec after the flash. Even with the clean system (D) of Fig. 10-4 with too low a termination resistor, severe waveform overshoot and some ringing are observed.

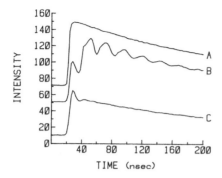

Fig. 10-6. Response of the different configurations of Fig. 10-4 to the laser excited decay of $[Ru(bpy)_3]^{2+}$ which has an ~400-nsec decay time. A corresponds to Fig. 10-4a, B corresponds to Fig. 10-4c, and C corresponds to Fig. 10-4d.

F. ELECTROMAGNETIC INTERFERENCE (EMI)

EMI, also known as radiofrequency interference (RFI), is one of the great nemeses of nanosecond fluorometry. Pulsed flash lamps and many pulsed lasers are great sources of EMI. Fast detectors and electronics, especially those used in single-photon counting systems, are exceedingly sensitive to these sources of noise. Indeed, the problem of reducing such noise to acceptable levels has consumed incredible amounts of research time and in some cases brought about the demise of an entire project—in particular for workers who have tried to build single-photon counting systems.

The subject of EMI is extremely diverse, and the interested reader is referred to the excellent book by Ott (1976). This book discusses in a lucid and readable form noise sources, propagation, reception, and suppression. The reader should be forwarned, however, that there are generally no simple solutions to noise problems. Only intelligent understanding of how noise escapes, enters, and affects your system and then hard work and luck can be counted on (usually) to solve these problems.

Probably the most important point is that noise sources such as flash lamps are radio and microwave transmitters. Given the tiniest opportunity the EMI will leak out of the source containers. The great brouhaha over microwave ovens is caused by the ease with which microwave radiation leaks from these ovens. Even a well-designed oven can leak milliwatts of radiation and still be within federal compliance. Such leakage from a laser could totally overwhelm many ancillary pieces of monitoring equipment. For example, a 2×0.1 in. crack (an open area of 0.2 in.^{2}) in the door or case of a flash lamp housing functions as a slot antenna and radiates enormous amounts of EMI. On the other hand a 2×2 in. square hole covered by a fine copper screen with a total open area of 2 in^2

radiates only a fraction of the EMI that the crack does. The reason is that the distance between conducting points on the screen is short compared to the wavelength of the EMI, which shorts the noise out and prevents its escape. An inspection of the glass window of a microwave oven reveals the metal screen used to block leakage through the glass.

The following rules are useful when considering the elimination of EMI.

(1) Faster signal sources are noisier and harder to eliminate the noise from than are slower ones. Therefore, use the slowest sources and electronics commensurate with your requirements. For example, a 1-nsec, 10^9 photons per pulse, single-photon counting lamp is noisier than a 1-MW, 1-μsec flash lamp-pumped dye laser. Adding capacitance to a flash lamp and a damping resistor may reduce the initial pulse rise time and, thus, the noise.

(2) Analog detection systems are generally less sensitive and much less prone to noise pickup than are single-photon counting ones.

(3) Always try to eliminate noise at the source rather than trying to keep it out of circuits once it has escaped. Proper flash lamp housing design and power supply connections to the lamp are one of the most important aspects of noise suppression in SPC systems.

(4) Build all noise sources into metal boxes. A black plexiglass or wooden box is an excellent light shield but has no effect on noise radiation. Keep the total volume and opening area in any noise source (i.e., flash lamp) cases to a minimum. Avoid slots. Aluminum miniboxes are very poor EMI shields because of the great difficulty of keeping the seams touching continuously. Use rf gasketing (copper screen or commercial shielding materials) on all seams. Cover all large holes with fine copper screen and make sure it is in contact with the case all the way around. Be wary of the use of paint, especially in systems with EMI gasketing. Paint can insulate the surfaces which must be electrically contiguous to shield properly.

(5) Filter and bypass all sensitive electronics and power connections entering the lamp housing. Ferrite beads are useful and inexpensive. Use EMI filters on all critical power lines.

(6) Use a single-point grounding scheme in any laser or flash lamp system (See Ott).

(7) Use EMI-shielded PMT housings in any SPC system and in laser systems if noise persists.

(8) Add an additional Faraday cage (a completely enclosed metal screen case) around the noise source. Connect the Faraday cage to a single-point ground. Power cables going to lasers or flash lamps are fre-

quent noise radiators. Filters at the source are frequently imperfect, and noise gets back into the cables. Even if they are already shielded, add another shield around them with a single-point ground. Car battery braid can be spread open and is quite useful for slipping over cables.

(9) With lasers or flash lamps equipped with a flowing gas system, the gas may become sufficiently conductive from the discharge to conduct EMI out of the enclosed area and then radiate it into the room. It is then necessary to put a highly conductive grounded trap on the exit side of the gas flow.

To demonstrate the subtlety of noise suppression problems, I relate a rather embarrassing but instructive incident from my laboratory. We had a small commercial spark gap-triggered nitrogen laser. This 1-kW laser had a 3-nsec pulse width, which meant that the spark gap rise time for 20 kV was probably subnanosecond. This extremely fast switching meant that the system was an incredible source of EMI. The manufacturer sold the system unshielded; it was the buyer's responsibility to provide as much shielding as required.

Attempts to use the unshielded laser with a sampling oscilloscope data acquisition system failed; the noise exceeded 100 mV for >500 nsec following the pulse. Even removing the laser to another room and firing it through an open door across 10 m of laboratory and into the sample gained negligible improvement. Trying an RFI filter on the laser power cord and putting the laser in a rather open metal cabinet yielded barely noticeable improvement. The laser was then situated in the next laboratory which had a double steel wall separating the detection electronics from the laser, and a small hole was drilled through the wall. Surely this shield would solve the problem; but, again, there was no great improvement. The steel plates were strapped together and connected to a 6-ft rod driven into salt-saturated earth—a rather major undertaking considering that the 8-in. reinforced concrete floor had to be drilled through first. The interconnecting laboratory door was shielded and grounded, with still no appreciable improvement.

A copper screen Faraday cage was built, and the entire laser and power supply were placed in it. There was some, but not enough, improvement (~70–100 mV of noise). Even opening the Faraday cage door had no noticeable effect on the noise level. Various grounding schemes were tried but to no avail.

Finally, a shielded EMI filter was added to the Faraday cage. When the laser was plugged into this filter in the cage, the noise level dropped by two orders of magnitude, and the system became usable.

In retrospect the problem was clear. The EMI filter was largely without effect initially because the noise was being transmitted directly from the laser and cabling. Later, when the Faraday cage was added, the noise was carried out of the cage on the power cord. The EMI passed through the shielded wall by following the wall wiring and came out through the plugs in the next room where it was either radiated or went directly down the power cords and into the instruments.

In summary, noise is extremely odious and difficult to avoid—especially in do-it-yourself projects. Many manufacturers of lasers and SPC equipment have had extensive experience, and prepackaged systems may avoid many problems and accelerate bringing a system on line. There have been notable exceptions in both laser and SPC instrument areas, however, and a purchaser of a commercial system is well advised to check with current users of the proposed systems, evaluate a system in his or her own laboratory, and obtain a written guarantee that the system will perform. This last point is particularly important if the user plans to assemble components from several different manufacturers.

G. SELF-ABSORPTION ERRORS

Most lifetime measurements assume that once a molecule has emitted a photon excitation is lost from the sample. If, however, the concentrations are high, self-absorption is large, and the escape path lengths are long, many of the emitted photons will be reabsorbed by the sample to produce other excited molecules with the same lifetime as the original one. When the photon yield is high, the excitation can then persist in the media by repeated emission–absorption–reemissions for periods far longer than the normal sample lifetime.

This reabsorption–reemission phenomena can produce serious errors in lifetime measurements if one is attempting to evaluate the actual molecular lifetime. For example, Wright (1955) found the decay time of a large anthracene crystal to be 18 nsec, whereas fine crystals which were largely free of self-absorption yielded a lifetime of 6.4 nsec. The discrepancy arises from the large reabsorption reemission in the single crystal. Concentrated solutions of high-yield materials with large emission absorption overlaps likewise suffer from the same problems. Many organic dyes, in particular, suffer from these problems. We stress that reabsorption–reemission errors are present in both emission and absorption methods of measuring lifetimes.

Hammond (1979) has studied concentration quenching of rhodamine dyes. The low degree of self-quenching necessitated high dye concentrations resulting in potentially serious errors. Hammond avoided these problems by means of a very thin-film (0.01 mm) cell which almost eliminated self-absorption. This type of cell should be invaluable as a general method for eliminating self-absorption reemission errors. Readers are referred to the original article for details and a complete mathematical analysis.

H. COMPENSATION FOR VARIATIONS IN FLASH PROFILE

In experiments that require deconvolution and long periods of data acquisition, the stability of the flash profile is of critical importance. Single-photon counting and boxcar averaging techniques are particularly prone to instability errors since recording of the decay and flash may require tens of minutes to many hours. During this extended time, if the excitation profile drifts between the measurement of $E(t)$ and $D(t)$, serious errors can result.

A logical way to minimize these problems is to measure $E(t)$ before and after the acquisition of $D(t)$ and use the average of the two for $E(t)$. Hazan et al. (1974) have carried this approach to its logical conclusion. They have developed an automatic sample changer shutter system which periodically makes measurements of $E(t)$, $D(t)$, and the background. In this way the collection of $E(t)$ and $D(t)$ transients are interleaved and one, in effect, collects and averages both $E(t)$ and $D(t)$ over the entire period. Their system is electronically and mechanically straightforward and can be readily adapted for single-photon counting experiments and other averaging techniques using a computer or multichannel analyzer (see also Easter et al., 1976).

We now describe the experimental procedure in greater detail. A single measurement cycle is divided into four portions: (1) The sample is inserted into the beam, the excitation shutter is opened, and a fixed number of averages are accumulated in one half of the MCA or the computer. (2) The shutter is closed, and then the same number of background sweeps are collected and subtracted from the $D(t)$ half of the MCA. (3) The sample is replaced with a scatterer or reflector, the shutter is opened, and a fixed number of $E(t)$ sweeps are added in the second half of the MCA. (4) The shutter is closed, and the same number of background sweeps are subtracted from the $E(t)$ half of the MCA. One cycle is now completed. The process is repeated as many times as necessary to obtain the desired signal-to-noise ratio.

The dark background readings are required to eliminate phototube dark current. More importantly, they also eliminate systematic instrument noise such as electromagnetic interference, amplifier baseline drift, and triggering shifts.

Figure 10-7 shows the results obtained without (Fig. 10-7a) and with (Fig. 10-7b) the drift compensation system. The severe nonrandom distortion associated with the uncompensated experiment, and the great improvements with the compensation system, are clearly demonstrated. Not only is the fit far better for the compensated data, but statistical analyses show these data to be fit well by a single-exponential impulse response. The uncompensated data, however, indicated incorrectly the presence of a second fast exponential.

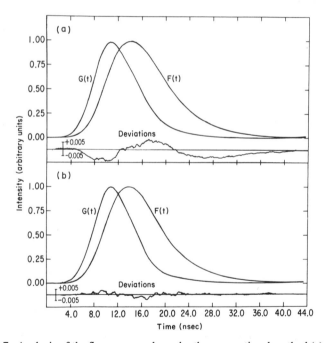

Fig. 10-7. Analysis of the fluorescence decay by the conventional method (a) and with a flash drift corrector (b) in which alternate sample and flash measurements are made. $G(t)$ is the excitation source and $F(t)$ is the fluorescence decay. The residuals following the least squares fit are plotted below each set of data. The greatly improved quality of the drift correction circuit is clearly evident. [From Hazan *et al.* (1974).]

We now demonstrate that this compensation scheme does, in fact, eliminate drift in $E(t)$, as it is by no means obvious that deconvolution will be successful if $E(t)$ changes shape (not just amplitude) during the course of data acquisition. The derivation follows that of Hazan *et al.* We will assume that we collect k separate background-corrected $D(t)$'s [$D_1(t)$, $D_2(t)$, . . . , $D_k(t)$] interleaved with k measurements of background-corrected $E(t)$'s [$E_1(t)$, $E_2(t)$, . . . , $E_k(t)$]. Further, we shall assume that the period of sampling is short enough so that the acquired $E_i(t)$ will be the appropriate one for $D_i(t)$. The final $E(t)$ and $D(t)$ are just the summation of each of the individual contributions. Thus:

$$E(t) = \sum_{j=1}^{k} E_j(t) \tag{10-16a}$$

$$D(t) = \sum_{j=1}^{k} D_j(t) \tag{10-16b}$$

If the impulse response of the sample is $i(t)$, then by the convolution integral

$$D_1(t) = \int_0^t i(t - u)E_1(u) \, du$$

$$D_2(t) = \int_0^t i(t - u)E_2(u) \, du$$

$$\vdots \tag{10-17}$$

$$D_k(t) = \int_0^t i(t - u)E_k(u) \, du$$

Combining the terms of Eq. (10-17) yields

$$\sum_{j=1}^{k} D_j(t) = \sum_{j=1}^{k} \int_0^t i(t - u)E_j(u) \, du$$

$$= \int_0^t i(t - u) \sum_{j=1}^{k} E_j(u) \, du \tag{10-18}$$

Using Eq. (10-16) we obtain

$$D(t) = \int_0^t i(t - u)E(u) \, du \tag{10-19}$$

which just states that the averaged $D(t)$ and $E(t)$ are related by a normal convolution integral and standard deconvolution methods applied to these quantities will yield $i(t)$ even though the shapes of $E(t)$ and $D(t)$ change during the acquisition.

I. TRIGGERING PROBLEMS

For the newcomer who does not fully understand the operation of an oscilloscope or transient recorder, their "internal" and "external" triggering characteristics can present another dangerous pitfall. Most deconvolution techniques depend on there being no time shift between $E(t)$ and $D(t)$, and improper selection of the system triggering mode can lead to a serious failure of this assumption.

In the internal triggering mode, the signal to be observed functions as its own trigger signal. The electronics sense when the signal passes a certain voltage going in a specified direction (i.e., positive or negative slope). Both slope and triggering level are adjustable. When the specified trigger conditions for the input signal are met, the oscilloscope sweep starts or data digitization begins.

External triggering works in the same way, except that the trigger signal is time-correlated with the waveform of interest but not directly derived from it. For example, a second PMT that views the excitation flash directly may be used to generate this trigger waveform (Fig. 2-1). Thus, once the triggering is set, the oscilloscope sweep will always start at the same time in relation to initiation of the event regardless of the characteristics of the main signal. This fixed starting time means that there is generally no timing problem in the external triggering mode.

To see how problems arise in the internal triggering mode, see Fig. 10-8. Typical noise-free $E(t)$ and $D(t)$ curves are shown in proper time relationships to each other in Fig. 10-8a. The lifetime used is 5.00 nsec. The triggering level, which has been set at 15% of full scale with a posi-

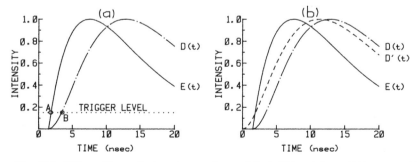

Fig. 10-8. Effect of internal triggering on observed flash $E(t)$ and decay $D(t)$ curves. (a) Properly time-related $E(t)$ and $D(t)$ showing the respective points at which each triggers the oscilloscope (A and B) for a 15% of full-scale positive slope triggering. (b) The observed time relationship between $E(t)$ and the apparent $D(t)$ denoted by $D'(t)$. The correct time relationship is given by $E(t)$ and $D(t)$ for comparison.

tive-going slope, is also shown. The initiation of the sweep for $E(t)$ is at point A, and the initiation of the sweep for $D(t)$ for the same triggering conditions is at point B. Thus, initiation of the two oscilloscope sweeps begins at different times relative to the true zero. Figure 10-8b shows the relationship of the two waveforms as they would be seen on the oscilloscope using internal triggering. The internally triggered $D(t)$ is denoted by $D'(t)$. The correctly timed decay is, again, denoted by $D(t)$. There is an ~1.5-nsec time shift between $D'(t)$ and $D(t)$. This shift, which does not really look large in comparison to the 20-nsec observation period, translates into a large error in deconvolutions.

We have used the phase plane method in an attempt to deconvolute the data of Fig. 10-8b using the properly timed $E(t)$ and $D(t)$ as well as using $E(t)$ and $D'(t)$. The corresponding phase plane plots are shown in Fig. 10-9. As expected, the plot derived using $E(t)$ and $D(t)$ is beautifully linear with a calculated τ of 4.995 nsec (fit over the 10–50-nsec interval). The plot using $D'(t)$ is disastrously nonlinear, even though the display begins at $t = 4$ nsec and only begins to approach linearity at 10 nsec [$W(t) = 0.14$]. A fit over the 10–50-nsec region yields a lifetime of 3.5 nsec or a 30% error.

If the oscilloscope or digitizer actually did begin to display or collect data at the trigger point, it would usually be immediately obvious that there was a problem. Both the flash and decay curves would appear to start at the same nonzero value. In the current example the apparent beginning of $E(t)$ and $D'(t)$ on the display would be at ~1.5 nsec in Fig. 10-8b.

Once noticed, corrective action such as adding a second PMT to view the flash, coupled with external triggering, would solve the problem. But

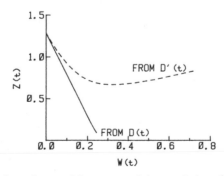

Fig. 10-9. Phase plane plots used for attempted deconvolution of the data of Fig. 10-8. The solid line is calculated from the properly time-correlated $E(t)$ and $D(t)$. The dashed line is calculated from the improperly related $E(t)$ and $D'(t)$. For the $D'(t)$ data, the point at $W(t) = 0.7$ corresponds to ~4 nsec in Fig. 10-8b.

the problem can be more insidious. Many oscilloscopes have internal delay lines. In the internal triggering mode, the signal triggers the oscilloscope and then passes through a long cable which delays its appearance at the display. This delay permits the display to show portions of the waveform that actually occurred before the trigger level was crossed. Thus, the full display of $E(t)$ and $D'(t)$ could be observed as shown in Fig. 10-8b. Again, close inspection of the data shown would clearly detect the problem; $D'(t)$ actually starts before $E(t)$ when internal triggering is used. However, the very fast rise of $E(t)$ in this case emphasizes the error. With slowly rising flashes and broad oscilloscope traces, it is frequently not possible to detect these errors visually, but the errors are still propagated to the final values.

With oscilloscopes one notices only the apparent pretrigger effect of the delay line in the short (nanosecond) time domain, since it is not possible to delay much longer than this with a physically reasonable cable. Thus, for slow (micro- to millisecond) phenomena viewed on an oscilloscope with internal triggering, one observes the abrupt rise and common starting point of both the flash and decay. Many transient recorders have a pretrigger mode that permits capture of the full waveform even at slow speeds and when much of it appears before the trigger. Thus, with these devices, as with oscilloscopes on faster sweeps, it is possible to conceal the internal triggering problem.

In summary, in measurements where temporal relationships between waveforms are critical, it is important to use external triggering rather than internal triggering of the display system. Similarly, if deconvolutions are to be performed, either external triggering must be used or a deconvolution method free of errors from time shifts between $E(t)$ and $D(t)$ must be employed (e.g., moments method 2). Finally, single-photon counting systems are, by their very nature, equivalent to externally triggered systems, since a PMT or an antenna that picks up EMI is always used to start the time-to-amplitude converter.

J. PILEUP PROBLEMS IN SINGLE-PHOTON COUNTING

Single-photon counting measurements have a unique form of distortion. This problem is pulse pileup. Pileup problems can arise from several different sources, and we will briefly describe these sources and ways to circumvent them. A lucid discussion of pileup problems with references is given by Harris and Selinger (1979), and the interested reader should consult this work. Mentioned, but not evaluated, is the mathematical method of Coates (1968, 1972; Bridgett *et al.,* 1970), which in principle can work to high-count rates.

The most obvious pileup problem arises if the counting rate is too high. A fundamental assumption of the SPC method is that no more than one stop photon occurs during the sweep of the time-to-amplitude converter. If this assumption is incorrect, then one no longer generates a proper histogram representative of the waveform because the TAC stops with detection of the first photon. If more than one counting pulse occurs during the TAC interval, only the earliest one will be detected and will produce a contribution to the MCA contents. In other words, short-time information will be acquired in preference to long-time information, and the early portion of the transient will grow in faster than it should relative to the later portions. This results in distortion of the waveform, which causes it to fall off more rapidly than it should. In the case of exponential decays .aken at too high a counting rate, the measured lifetimes will be too short (vide infra).

For purposes of demonstration, we consider a single-exponential decay for the sample given by $\exp(-t/\tau)$, although Harris and Selinger (1979) also give results for a constant-intensity source and for the difference and the sum of exponentials. For the single-exponential decay measured on an SPC instrument, the observed impulse response will be given by (Harris and Selinger, 1979)

$$N(t) = A \exp\{-u[1 - \exp(-t/\tau)]\}[\exp(-t/\tau)] \qquad (10\text{-}20a)$$

$$u = n_0\tau \qquad (10\text{-}20b)$$

where $N(t)$ is the mean number of counts in the channel corresponding to time t, A is a proportionality constant, and u is the mean number of counts received at the TAC following a flash. For a system running with a 10-kHz lamp, u will be 0.1 if 1000 stop pulses per second were detected by the stop phototube. This expression shows that, for $u > 0$, $N(t)$ will not be purely exponential since the preexponential factor for the purely exponential decay term is time dependent. Only as $u \rightarrow 0$ does the preexponential factor approach being time independent.

Figure 10-10 shows the expected observed decay curves for a single-exponential decay measured with different values of u. The presented decays are noise-free to prevent cluttering of the plots. Real data would, of course, exhibit Poisson noise. Harris and Selinger give experimental curves for u's up to 0.7 and obtain good theoretical agreement. The plot for $u = 0.01$ is indistinguishable within a pen's width for the curve for $u = 0$. For $u > 0.01$, however, very appreciable distortions in the curves arise, and it is clear that attempts to fit these curves to a single exponential

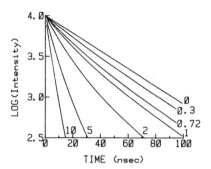

Fig. 10-10. The effect of excessive pulse count rates in monitoring a simple exponential decay using a single-photon counting instrument. The lifetime is 40 nsec. The number indicated by each curve is the average number of photons that would be detected by the stop photomultiplier tube per flash. An average value of 0.01 is indistinguishable within a pen's width of the 0 value. The curves are calculated and presented without noise to emphasize the relationship of waveforms.

would result in erroneous lifetimes and poor fits. Indeed, the standard way to avoid pileup errors of this type is to keep u small—typically <0.05 and frequently less than 0.01. Of particular note are the waveforms for large u's (e.g., 5–10). In these cases the apparent exponential decay is shortened enormously. This phenomena is well known to users of single-photon counting instruments. If the light level is too high, the observed pulses can be made artificially as short as desired.

Harris and Selinger (1979) reported experimental measurements on the exponential decay of pyrene for different values of μ where μ was the mean number of events per TAC cycle. They found calculated lifetimes to be good (better than 1%) and with statistically acceptable χ_r^2 for μ's up to 0.03. For larger μ's, the χ_r^2 increased and the accuracy of the measured lifetimes decreased rapidly. For example, at $\mu = 0.22$, they found τ to be 12% low; at $\mu = 0.32$, τ was 16% low; at $\mu = 0.50$, τ was 23% low, and at $\mu = 0.70$, τ was 30% low. They concluded that, for accuracy better than 2% in simple exponential fits, μ must be <0.03.

Harris and Selinger (1979) also proposed an alternative approach to pileup corrections of this type. They merely used an expression analogous to Eq. (10-20) as the sample impulse response and carried out a nonlinear least squares fit with this function. Using this expression, they found the statistical fits for the pyrene data to be acceptable up to $\mu = 0.70$, although above $\mu = 0.50$ there was a small increase ($\sim$2–3%) in the calculated τ. These authors attribute this to the fact that the equation is nonlinear in μ. If the flash intensity fluctuates, then the μ for each flash fluctuates. They conclude that, for 1% maximum error using this nonlinear fit, μ's up to 0.1 could be used if the source did not fluctuate by more than 10%. This nonlinear fitting scheme is, of course, applicable to sums and differences of exponential functions where Harris and Selinger give the necessary fitting equations. One disadvantage of this approach is that, if deconvolutions of τ's from both $E(t)$ and $D(t)$ are required, there is no fast convolution integral formula available for their equation. Thus, fitting by these approaches would be much slower.

This statistical problem can also be treated experimentally by means of a device called a pileup inspector (PUI) (Davis and King, 1970; Knight and Selinger, 1973). A typical PUI inhibits storage of the TAC output by a time greater than the TAC range. If no additional pulse is detected from the stop pulse to the end of this time-out period, then there will be no pileup and the signal will be accepted and processed by the MCA. When any additional pulses occur, a pileup condition exists and the TAC output is not sent to the MCA. In the absence of system dead time errors (vide infra), such a PUI scheme works well up to count rates per TAC cycle of 0.37 which corresponds to $u = 1$.

A second type of pileup is associated with the finite resolving time of the detection system. Every detector has a dead time T after it detects a pulse and before it can detect another one. For an intriguing example of dead time, Harris and Selinger (1979) have analyzed the original α-particle counting experiments of Rutherford and Geiger (1910) and have arrived at a dead time of 0.53 sec for the human eye; the apparent statistical unacceptability of the original Rutherford–Geiger data is then traced to a failure to recognize the finite T and the high counting rates of 0.53 counts/sec in the original experiments.

In SPC counting experiments, the finite T has the undesirable effect of hiding counts. Thus, if one is judging the number of photons detected per TAC cycle merely by taking the ratio of stop pulses to the flash rate, one could be substantially in error if the intensity is so high as to run pulses together. Further, the successful operation of a PUI depends on being able to resolve additional pulses from the stop pulse. However, if an additional pulse comes through during a period T after the stop pulse, the extra pulse is missed; then a TAC cycle is accepted that the PUI should have rejected. This failure of the PUI becomes especially serious for short lifetimes where the TAC range is short, since T can be on the order of 10 nsec. In the rather extreme case of a TAC range of 10 nsec and $T = 10$ nsec, one could never detect pileup at all by means of a PUI.

Energy discrimination schemes have been developed that can detect pulse pileup when the pulse comes through closer than T. This technique uses the fact that some PMTs have different output pulse heights depending on whether one, two, or three photons were detected. As expected, a one-photon output spike has, on the average, a much lower peak height than two- or three-photon spikes. Thus, by monitoring the output pulse by a single-channel analyzer, one can discard all pulses that are too large or too small to correspond to a well-defined one-photon output pulse. This discrimination is not perfect, since there is a finite probability of a two-photon pulse having a height within the window of acceptance for a one-photon pulse. This subject, including selection of the window, is discussed in detail by Kouyama (1978).

The limitation of energy discrimination in correction of dead time pulse pileup means that this approach will work best for short lifetimes where a simple PUI begins to fail. Harris and Selinger (1979) have shown that, for systems with $T = 10$ nsec, a simple PUI is best for lifetimes greater than 80 nsec, whereas energy discrimination is better for shorter lifetimes.

K. PURITY PROBLEMS

The sensitivity and wide dynamic range of modern decay time instruments is so great that it is easy to lose sight of the fact that what is being measured may be an impurity rather than the system of interest. It is, for example, easy to see and measure decay kinetics on a highly luminescent component which may represent only a few parts-per-million impurity in a nonluminescent material.

Further, even a few parts per thousand of a long-lived impurity can easily distort the trailing edge of a short-lived main component (see, for example, Fig. 5-8). Similarly, a short-lived impurity can distort the early times. The three and four decades of good fit to decay curves seen in the manufacturer's literature and in some research publications are obtained with carefully selected and purifed systems. The average researcher is either unable to get his chemical systems so clean and well behaved or does not have to have it this clean to answer the questions of interest. For example, the author heard a biochemist describe one of his encounters with an SPC instrument. This chemist had spent years purifying a protein which, in his area, was exceptionally pure (>95%). But when the decay curve was measured, all information from the main protein appeared to be washed out by the impurity contributions after only the first decade of the signal decay.

Thus, anyone using luminescent lifetime measurements should make measurements on materials of known quantum yields that emit in the region of interest. It is then possible to estimate what impurity concentrations and quantum yields can produce specific signals or distortions.

Further, it is important to recognize that glasses and solids can produce nonexponential decays even with pure materials. Different sites can exist in rigid glasses as well as different crystallographic sites in crystalline materials; molecules at these different sites can exhibit different lifetimes either because of differing environments or because of energy transfer to other molecules. Solids can be particularly insidious as to impurity emissions; energy transfer through the lattice can efficiently excite trace impurities.

Finally, particularly with SPC instruments, workers invariably run into systems that exhibit less than ideal decay kinetics. It then becomes essential to analyze carefully one's goals with the system. Is it essential to have the system absolutely pure, which may require years of work or never be achieved at all? Is some reasonable enhancement of the purity adequate? Are the current data adequate to answer the questions concerning the system? Are the answers to additional questions worth the necessary work involved in purifying the system? If one anticipates these questions before making any measurements, then one will frequently avoid time-consuming side trips.

Testing and Evaluation of Methods and Instruments

A. INTRODUCTION

Verification of the satisfactory performance of an instrument, a new data reduction technique, or a new reduction program is a crucial step in the operation of any system. This can be done in two ways. Compounds of known lifetimes can be measured, and the calculated decay times compared with the accepted ones. Alternatively, synthetic data can be generated and then the decay parameters calculated using the desired algorithm and program. These results are then compared with the original parameters. Each approach has an important place, and generally both methods are used to complement each other.

The chemical standards method has the advantage of testing the entire system including the hardware and software. It suffers from the following disadvantages:

(1) The number of accurately known standards, especially with very short lifetimes, is relatively limited.

(2) The testing of many different data sets can become extremely time-consuming, especially with a slow data acquisition technique such as single-photon counting.

(3)   Standards may be difficult to use because of such problems as impurities, oxygen quenching, and temperature effects. Thus, if the "correct" lifetimes are not obtained, it may not prove that the system is working incorrectly.

Simulation techniques, on the other hand, cannot test the system hardware. They can check only reduction algorithms and software. They are, however, very fast, and it is a simple matter to verify software performance over a wide range of "experimental" conditions, even though chemical standards with these values may not be available.

In this chapter we discuss the technique of digital simulation including methods of selecting suitable curves as well as generating and adding noise. The important problem of verifying that the noise is indeed "good" noise is covered at length. Finally, we describe chemical systems suitable as standards.

B. DIGITAL SIMULATIONS

A number of different algorithms for data reduction of luminescent transient phenomena are presented in this volume. The limited number of highly accurate and reliable reference systems makes it virtually impossible to evaluate the precision and accuracy of a data reduction approach experimentally. Further, it is experimentally difficult to evaluate conveniently and systematically the effects of variations in such parameters as noise, ratios of different components, and relative lifetimes of these components on a given method.

To avoid these problems many workers resort to digital simulations in which synthetic data are generated that mimic real data as closely as possible. The data reduction algorithm is then applied repeatedly to a series of these synthetic data sets. Since the sample parameters that went into the calculation of the synthetic data are known, the accuracy and precision of the data reduction scheme can be evaluated without ever having to do a single experiment. Further, since the simulation parameters are readily varied, it is a simple matter to evaluate accurately the expected performance and reliability of the approach. As an added benefit many data reduction techniques, especially deconvolutions of multiple exponentials, work best when the operator applies them in certain ways (e.g., in determining what fraction of the curve to fit or the degree of exponential depression). Thus, digital simulation provides an ideal way for the operator to develop skill and confidence in a specific method.

For successful simulation a number of conditions, including several subtle ones, must be satisfied. Reasonably accurate noise statistics for the data must be defined, and realistic noise-free $E(t)$'s must be available. Further, computation of noise-free $D(t)$'s must not introduce any systematic errors in $D(t)$. The noise added to $E(t)$ and $D(t)$ must follow faithfully the expected experimental noise statistics. We will discuss the problems of selection of $E(t)$ and suitable noise statistics as well as methods and pitfalls in the generation of suitable $D(t)$'s. Methods of generating suitable noise distributions, and particularly the problems of verifying that the synthetic noise is indeed reliable, are covered later.

1. Selection of $E(t)$

Several analytical functions have been widely used for simulations. These include:

$$E(t) = \exp(-t/A) - \exp(-t/B), \quad A > B \tag{11-1}$$

$$E(t) = t^2 \exp(-t/D) \tag{11-2}$$

Both functions are zero at $t = 0$, rise rapidly to a maximum, decay off rapidly and, at long enough times, are exponential. Our own preference is the first, and all simulations carried out for this book used it. The shape is easily maintained while varying the width merely by changing A and keeping A/B constant. The rising edge can be readily made very rapid by reducing A, and the decay time at long times can be largely controlled by B. Equation (11-1) is also especially easy to use for analytical evaluation of $D(t)$ which is a sum of exponentials. Both functions, however, lack the long, low-intensity tail characteristic of many real flashes. If necessary, it is possible to simulate this tail by making $E(t)$ a properly weighted sum of two terms of the form of Eq. (11-1) or (11-2); the second term would have longer lifetimes than the first.

Many workers use an experimentally determined $E(t)$. This is usually done with single-photon counting where the necessary several orders of magnitude in intensity can be accurately acquired by collecting enough counts in each channel. It is essential, however, to acquire a good, virtually noise-free $E(t)$, because only then can noise be added to it in a reliable fashion.

2. Generation of $D(t)$

$D(t)$ can be generated by numerical convolution of the discrete representations of $E(t)$ with the impulse response $i(t)$ of the proposed decay function. Alternatively, $D(t)$ may be generated analytically from a functional $E(t)$ [e.g., Eq. (11-1) or (11-2)], an analytical $i(t)$, and the convolution integral.

Numerical convolution using the Grinwald-Steinberg iterative convolution formula [Eq. (6-20)] is fast, requires no functional form for $E(t)$, and works conveniently with any impulse response function that is a sum of exponentials. Its primary disadvantage is that it is not exact; a numerical trapezoidal integration rule is used, and errors are inherently introduced into the calculated $D(t)$. These errors are most serious when the density of points is low and the τ's of $i(t)$ are short. See Section 6.D for representative errors. The simplest way to check for errors of this type is to deconvolute the decay parameters from the noise-free $E(t)$ and the calculated $D(t)$ by using an established method (e.g., the phase plane method for $i(t)$ $\propto \exp(-t/\tau)$. Virtually the entire range of $E(t)$ and $D(t)$ should be used. The deconvolution method chosen must not use the same convolution approach as that used to generate $D(t)$.

As an example of this problem we used $E(t)$ of the functional form of Eq. (11-1) with $A = 5$ nsec and $B = 6$ nsec. The resultant $E(t)$ exhibited an ~14-nsec FWHM. Using a data point every nanosecond, $D(t)$ was generated by the GS formula. Deconvolutions of noise-free data by the phase plane method for single-exponential $i(t)$'s consistently yielded low estimates of τ with errors amounting to several percent for shorter lifetimes. A finer grid of times would reduce the errors from this source at the expense of increased computational time.

Demas and Crosby (1970) have also described an iterative convolution method that minimized these problems. The GS formula evaluates the integrand of Eq. (6-20) at discrete points and then uses a trapezoidal rule numerical integration of the pointwise function. Demas and Crosby assumed that $E(t)$ could be represented by straight-line segments between $E(t)$ points and then carried out the integrations analytically and therefore exactly. The Demas–Crosby approach has greater accuracy (especially for shorter lifetimes) than the GS method at the expense of a more complex formula and far greater computational time. Thus, the GS formula, coupled with a higher density of data points, will generally be a better approach.

In the analytical approach $D(t)$ is evaluated directly from the analytical form of $i(t)$ and $E(t)$. This is done with essentially perfect accuracy regardless of how many points are used. For example, if $i(t) = \exp(-t/\tau)$ and $E(t)$ is given by Eq. (11-1), then $D(t)$ becomes

$$D(t) = \exp(-t/\tau)\{[\exp(Rt) - 1]/R - [\exp(St) - 1]/S\}$$
$$R = (A - \tau)/A, \qquad S = (B - \tau)/B$$

(11-3)

This formula is particularly convenient for simulations, since most of the important impulse responses are sums of exponentials. No errors arise in the calculated $D(t)$. The only inconvenience is that, if the time interval is too long compared to some of the τ's, then the ranges of the evaluation of $\exp(-t/\tau)$ in the computer may be exceeded and an under- or overflow may result. At the expense of slightly increased program complexity, this is readily overcome. Also, τ cannot equal A or B.

The analytical approach for generating $D(t)$ is somewhat slower than the GS approach. Generally, however, the generation of $D(t)$ is only a small fraction of the total time in a simulation. For example, many different noisy $E(t)$'s and $D(t)$'s are normally generated and reduced from a single noise-free $E(t)$ and $D(t)$.

3. Addition of Noise

The type of noise added to the $E(t)$ and $D(t)$ will depend upon the needs and resources of the investigator. If the data are to simulate single-photon counting statistics, then the $E(t)$ and $D(t)$ curves should be scaled to appropriate numbers of counts per channel. Commonly, the curves are scaled to peak counts of 10^3–10^6 with 10^4 being a popular value because of its wide use in actual experiments. Noise is then superimposed on both curves following a Poisson distribution. For high enough count numbers (>20) Poisson noise approximates closely a normal distribution with a standard deviation equal to the square root of the number of counts. Thus, if we have a Gaussian noise generator G which has a mean of zero and a standard deviation of 1, we replace each observed channel count C_i with noisy data:

$$^nC_i = C_i + G_i(C_i)^{1/2} \tag{11-4}$$

where ^nC_i is the noisy data in channel i. A new G_i is generated for every channel. We discuss the generation of a Gaussian G_i later in this chapter.

For channel counts much below ~20, a Poisson distribution deviates appreciably from a normal distribution; the Poisson distribution is limited to values ≥ 0, whereas Gaussian noise can range from $+\infty$ to $-\infty$. For large C_i's, however, the chances of ^nC_i becoming negative are vanishingly small. For example, if $C_i = 25$, its standard deviation will be 5. The chances of G_i assuming a value of -5, which would be required to make ^nC_i negative, are less than 0.00003% which could routinely be ignored. Many workers use Eq. (11-4) even for ^nC_i well below 25. Then, if an ^nC_i less than 0 is generated, ^nC_i is set equal to zero or the generation process is repeated until an acceptable ^nC_i is obtained. While not rigorously correct, this procedure is certainly acceptable if most of the C_i are >25 counts. If, however, one is simulating real single-photon counting data where a significant number of channels contain <20 counts, it becomes imperative to use a true Poisson noise generator. In this case, each C_i is replaced by the value generated from a Poisson noise generator P_i, which returns a value which averages to C_i and has a standard deviation of $C_i^{1/2}$. Such a generator is described later in this chapter.

The choice of noise statistics is much less clear if the results are to be compared with the data taken from an oscilloscope photograph or a transient digitizer. This noise can be approximated as Gaussian, but the scaling factor and its dependence on amplitude are less certain. Some workers assume a constant amplitude of noise, but this is likely to be true only if amplifier noise dominates. A reasonable way to evaluate the noise is to record different levels of signals of constant intensity and measure the standard deviation. However, this method does not take into account the variations of variable trace width and reader bias on reading data off a decay curve.

A less accurate, but not unreasonable approach, is to estimate the noise level at the largest signal amplitude, treat it as if it were Poisson, and calculate the channel counts C_i that would yield such a noise level. Then proceed as if the signal and noise were derived from single-photon counting.

The assumption is occasionally made that the noise is of uniform probability over a fixed range and zero outside this range. For simulation this is computationally simple, since most computers have random number generators that return a uniformly distributed random number over the range 0–1. At least to this author, however, this appears to be an unreasonably physical model. We do not recommend the use of a uniform noise generator, since it is so easy to generate more reasonable Gaussian or Poisson noise.

C. GENERATING SYNTHETIC NOISE

In digital simulations it is normal to use the computer's built-in random number generator to generate the requisite noise to add to the signals. The two problems that arise here concern the desired distribution of the noise and verification of the "quality" of the noise.

First, most computers do not generate noise of the correct functional form. We will describe ways in which other distributions may be derived from the common uniformly distributed generator.

The second problem is more subtle. Most so-called random number generators are, in fact, not random number generators at all. They are precisely defined sequences of numbers generated by a mathematical algorithm and, if the proper algorithm is used, the generated sequence will appear to be random. Such generated sequences are, in fact, properly called pseudorandom numbers and they can possess underlying patterns or structures that can introduce systematic errors into synthetic noisy data generated from them. Such bias can have two undesirable consequences. First, an algorithm that would have worked well on real data may fail completely or give poorer results on the synthetic data, which will result in discarding a perfectly acceptable approach. Conversely, the synthetic noise might have such regularity as to make an unacceptable data reduction technique appear to be useful. These observations are especially true since many pseudorandom number generators always reset to the same initial value and give exactly the same sequence every time a program is run. Thus, if the initial set of noise is an especially favorable or unfavorable distribution, an erroneous impression is created. Methods of evaluating the quality of noise for use in digital simulations will be discussed in Section D.

We shall now describe how to generate other noise distributions starting with the commonly available uniform random number generator. We assume that this noise generator returns random numbers U's which are uniformly distributed over the 0–1 interval. The following discussion is derived from Knuth (1969).

1. Gaussian Noise

We shall discuss and compare three common and readily programmed methods for generating random Gaussian noise from U. These methods are the Box and Muller (1958) algorithm, the modified Box, Muller, and Marsaglia algorithm, and Teichroew's algorithm.

The Box, Muller, and Marsaglia algorithm (Knuth, 1969) proceeds as follows: Generate two U's, U_1 and U_2, and then calculate V_1 and V_2:

$$V_1 = 2U_1 - 1 \qquad (11\text{-}5a)$$

$$V_2 = 2U_2 - 1 \qquad (11\text{-}5b)$$

V_1 and V_2 are uniformly distributed over the interval -1 to $+1$. Calculate S from

$$S = V_1^2 + V_2^2 \qquad (11\text{-}6)$$

If $S > 1$, a successful normally distributed random number will not be generated; new U_1 and U_2 are generated, and the process is repeated. If $S \leq 1$, compute X_1 and X_2 from

$$M = [(-2 \ln S)/S]^{1/2} \qquad (11\text{-}7a)$$

$$X_1 = V_1 M \qquad (11\text{-}7b)$$

$$X_2 = V_2 M \qquad (11\text{-}7c)$$

X_1 and X_2 are independent and normally distributed with a mean of 0 and a standard deviation of 1. The algorithm as described does not include a test for $S = 0$, which is unacceptable in Eq. (11-7a). The reason is that most pseudorandom number generators never give two succeeding values that are equal and, thus, S will never be 0. If another method of generating U's can yield $S = 0$, a test for $S = 0$ must be made and the generation sequence repeated if necessary. This algorithm is essentially perfect for generating normal deviations.

A simpler version of the Box and Muller (1958) algorithm is given by

$$R = 2 U_1 \qquad (11\text{-}8a)$$

$$T = (-2 \ln U_2)^{1/2} \qquad (11\text{-}8b)$$

$$X_1 = T \cos(R) \qquad (11\text{-}8c)$$

$$X_2 = T \sin(R) \qquad (11\text{-}8d)$$

This expression is essentially equivalent to Eq. (11-7) but is generally computationally slower.

Teichroew's method (Knuth, 1969) for generating normal deviations requires no transcendental functions. It is carried out as follows: Calculate 12 independent U's, U_1, U_2, . . . , U_{12}. Then calculate

$$R = \left[\left(\sum_{i=1}^{12} U_i \right) - 6 \right] \Big/ 4 \qquad (11\text{-}9)$$

Then the normal deviate X is calculated from

$$X = \{[(a_9 R^2 + a_7) R^2 + a_5 R^2 + a_3] R^2 + a_1 \} R \qquad (11\text{-}10)$$

where $a_1 = 3.949846138$, $a_3 = 0.252408784$, $a_5 = 0.076542912$, $a_7 = 0.008355968$, $a_9 = 0.029899776$. Here R is approximately normal, with a mean of 0 and a standard deviation of 0.2500. Equation (11-10) is a polynomial approximation which converts R into a more nearly Gaussian distribution. X's generated by Teichroew's method can never assume values outside the range ± 8.65. The probability of an X in the normal distribution falling outside this range, however, is almost nonexistent.

For computational speed the three approaches were tested on a Hewlett-Packard 9825A microcomputer. To generate 1000 normal deviates the first [Eqs. (11-6) and (11-7)] and second [Eq. (11-8)] Box and Muller approaches required 13 and 25 nsec, respectively, whereas Teichroew's method required 55 sec. The relative speeds may vary significantly from machine to machine depending on the speed required for different mathematical functions. From a programming standpoint, Teichroew's method requires no transcendental functions and the second Box and Muller method is more straightforward to program. In most instances the Box–Muller–Marsaglia algorithm is the preferred one.

2. Normal Distributions with Other Means and Standard Deviation

The two algorithms above generate a distribution with a mean of 0 and a standard deviation of 1. To generate a new distribution of Y's with a mean of μ and a standard deviation of σ, merely calculate the Y's from the X's as follows:

$$Y = \mu + \sigma X \qquad (11\text{-}11)$$

This expression merely shifts the origin of the X distribution and expands or contracts its width by the factor σ.

Figure 11-1 demonstrates the agreement between a normal distribution with a mean of 50 and a standard deviation of 15 and the simulated

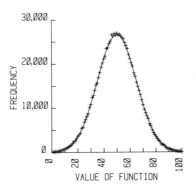

Fig. 11-1. Distribution of synthetic Gaussian noise generator with a mean of 50 and standard deviation of 15 (+) compared with the theoretical distribution (——). The synthetic distribution was generated by Box–Muller–Marsaglia algorithm and Eq. (11-5) to (11-7). Then 1,000,000 random deviates were generated, and the data are plotted as a histogram with each window being 1 unit wide and centered on integer values. The Gaussian curve has been normalized to the same area.

distribution generated from Eqs. (11-5) to (11-7) and Eq. (11-11). For the simulation 1,000,000 normal deviates were generated, and a histogram plot is displayed with a window width of 1 for each bin. The agreement between the observed and theoretical curves is quite good.

3. Poisson Noise

Poisson statistics must assume discrete values. If the average number of counts is μ, then the probability $P(x, \mu)$ of having x counts in a channel is given by (Bevington, 1969)

$$P(x, \mu) = [\mu^x \exp(-\mu)]/x! \qquad (11\text{-}12)$$

In all cases the standard deviation is given by $\mu^{1/2}$. Table 11-1 summarizes $P(x, \mu)$ for several different μ's. Particularly for low μ's there is great dissymmetry in the distribution and a high probability of finding zero counts. For example, with $\mu = 1$, there is an equal probability of finding 0 or 1 counts. For larger μ's, however, the distribution becomes more nearly symmetrical and very Gaussian-like.

An algorithm (Knuth, 1969) for generating Poisson noise is as follows:

(1) Initialize the variables:

$$N = Q = 0, \qquad P = e^{-\mu} \qquad (11\text{-}13)$$

(2) Generate the uniformly distributed U and calculate

$$Q = QU \qquad (11\text{-}14)$$

(3) If $Q > P$, then set

$$N = N + 1 \qquad (11\text{-}15)$$

and go back to step 2.

(4) The calculation is complete, and N is the random number having a mean of μ and following a Poisson distribution.

Table 11-1 contains several calculated distributions using the Knuth algorithm. In each case the deviation was obtained for 100,000 Poisson deviates calculated by the Knuth algorithm. The Knuth algorithm becomes increasingly slower as μ becomes larger, so it is likely to be unacceptably slow for $\mu > 20$. Fortunately, the Poisson distribution so closely mimics a Gaussian distribution for these high counts that the normal distribution, with a standard deviation equal to the square root of the channel counts, can be used for these large μ's.

TABLE 11-1

Distribution of Synthetic and True Poisson Noise

N	μ= 20.0 Calc	True	10.0 Calc	True	5.00 Calc	True	1.00 Calc	True
0	0.00000	0.00000	0.00009	0.00005	0.00620	0.00674	0.36803	0.36788
1	0.00000	0.00000	0.00020	0.00045	0.03415	0.03369	0.36730	0.36788
2	0.00000	0.00000	0.00215	0.00227	0.08415	0.08422	0.18611	0.18394
3	0.00018	0.00000	0.00745	0.00757	0.13992	0.14037	0.05947	0.06131
4	0.00000	0.00001	0.01603	0.01892	0.17679	0.17547	0.01543	0.01533
5	0.00036	0.00005	0.03808	0.03783	0.17611	0.17547	0.00295	0.00307
6	0.00018	0.00018	0.06124	0.06306	0.14607	0.14622	0.00067	0.00051
7	0.00036	0.00052	0.09190	0.09008	0.10719	0.10444	0.00003	0.00007
8	0.00162	0.00131	0.11595	0.11260	0.06363	0.06528	0.00000	0.00001
9	0.00251	0.00291	0.13204	0.12511	0.03426	0.03627	0.00001	0.00000
10	0.00643	0.00582	0.12374	0.12511	0.01769	0.01813		
11	0.01093	0.01058	0.10986	0.11374	0.00874	0.00824		
12	0.01809	0.01763	0.09311	0.09478	0.00340	0.00343		
13	0.02938	0.02712	0.07364	0.07291	0.00143	0.00132		
14	0.03545	0.03874	0.04779	0.05208	0.00000	0.00047		
15	0.05248	0.05165	0.03474	0.03472	0.00022	0.00016		
16	0.06304	0.06456	0.02217	0.02170	0.00000	0.00005		
17	0.07680	0.07595	0.01354	0.01276	0.00000	0.00001		
18	0.07881	0.08439	0.00752	0.00709				
19	0.09382	0.08884	0.00418	0.00373				
20	0.08754	0.08384	0.00270	0.00187				
21	0.08533	0.08461	0.00131	0.00089				
22	0.07483	0.07691	0.00038	0.00040				
23	0.06685	0.06688	0.00019	0.00018				
24	0.05989	0.05573	0.00000	0.00007				
25	0.04602	0.04459	0.00000	0.00003				
26	0.03439	0.03430	0.00000	0.00001				
27	0.02242	0.02541						
28	0.01629	0.01815						
29	0.01346	0.01252						
30	0.00950	0.00834						
31	0.00536	0.00538						
32	0.00268	0.00336						
33	0.00269	0.00204						
34	0.00071	0.00120						
35	0.00054	0.00069						
36	0.00071	0.00038						
37	0.00035	0.00021						
38	0.00000	0.00011						
39	0.00000	0.00006						
40	0.00000	0.00003						
41	0.00000	0.00001						
42	0.00000	0.00001						

Thus, for the most accurate simulations of single-photon counting statistics we recommend the following procedure. For $C_i > 20$ counts use the Gaussian distribution of Eq. (11-4). For $C_i \leq 20$ use the Knuth algorithm of Eqs. (11-13) to (11-15) to generate the Poisson data directly. The changeover from integer data for $C_i \leq 20$ to continuous data for $C_i > 20$ will be inconsequential in the simulations.

D. TESTING SYNTHETIC NOISE

There are innumerable tests for evaluating the quality of synthetic noise. The empirical tests described here can be divided into two basic classes: (a) distribution tests and (b) sequence tests. The number of tests discussed here may seem excessive, but the reader should be warned that a random number generator passing all the tests is not necessarily proven suitable for simulations. Each test is optimized for the detection of differing types of nonrandomness, but some patterns may still escape detection and make use of the generator unsuitable in certain simulations. The discussion of the chi-square, Kolmorogov–Smirnov (KS), and serial tests follow Knuth (1969). The discussion of runs tests is adapted from Gibbons (1976).

Visual inspection of the noise is, in general, a worthless test of the quality of the noise. Any noise distribution that would fail visually is an abominably bad generator and would certainly fail any of the standard tests given below. Figure 11-2 compares two uniform random number generators distributed over the interval 0–1. Two hundred calls to each

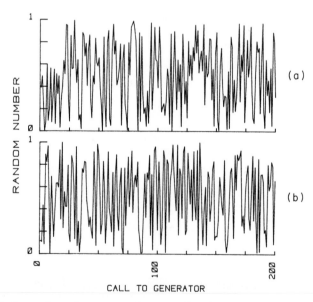

Fig. 11-2. Comparison of a very poor (generator 2) and a good noise generator (generator 1) with uniform probability distribution over the interval 0 to 1. The value of the random number is plotted versus the call to the random number generator. All the points are connected to aid visualization. For the moment it is left up to the reader to try to determine visually which generator is which.

generator have been made and plotted versus the call number. Visually, both look quite reasonable and noisy. However, one generator in Fig. 11-2 is acceptable by all tests, and one in Fig. 11-2 fails the mean and standard deviation test, the chi square test, the KS test, one runs test, and the serial test. In short, in spite of the visual acceptability of both generators, one is disastrous for simulations and the other is quite acceptable. For the moment it is left to the reader to try to determine which is the good generator. Complete testing of both will be described later.

1. Mean and Standard Deviation Test

Before beginning any of the more sophisticated tests described below, perhaps the simplest test for the standard uniform noise distribution between 0 and 1 is to evaluate the mean and standard deviation of the noise for a very large number of random numbers (e.g., 10^4–10^6). These values should be reasonably close to the predicted mean and standard deviation of 0.5 and 0.2886, respectively. Failure at this stage would indicate an exceptionally poor generator which should be discarded.

To quantitate what represents a failure of the mean test, we note that, if N quantities are averaged to yield the mean M and a standard deviation σ, then the standard deviation of the mean σ_m is $\sigma_m/N^{1/2}$. We assume that the expected mean is μ. Thus, when N values are averaged, we expect M to fall within $\mu \pm \sigma/N^{1/2}$ 68% of the time, within $\mu \pm 2(\sigma/N^{1/2})$ 95% of the time and within $\mu \pm 3(\sigma/N^{1/2})$ 99.7% of the time. This statement is correct if the noise is normally distributed. For a uniform generator, one will expect a somewhat different distribution, but an M falling outside $\mu \pm 5\sigma_m$ certainly would be unacceptable.

In a series of tests of 10,000 successive calls on a uniform generator, we found values of M and σ of 0.4627 ± 0.2783, 0.4624 ± 0.2836, and 0.4637 ± 0.2824. The expected value is 0.5000 ± 0.2886. The expected σ_m is 0.0028 (= 0.28/10,000$^{1/2}$). The M's all fall over $10\sigma_m$ away from the expected 0.5000, and this generator should not be used for precision work. Another generator for 10,000 successive calls gave M's and σ's of 0.5019 ± 0.2890, 0.5022 ± 0.2897, and 0.50011 ± 0.2889. All the averages are within $\pm\sigma_m$, and this generator was considered acceptable at this stage of the testing.

2. Chi-Square Test

We now consider distribution tests. The simplest of these is a box, bin, or histogram sort followed by a chi-square test to verify that the deviation from the expected distribution is acceptable. Assume that we have k categories into which an observation can fall and we make N independent determinations. Let p_s equal the theoretical probability that the observation will fall in the sth category and Y the actual number of measurements that fall into that category. Then χ^2 is given by

$$\chi^2 = \sum_{s=1}^{k} (Y_s - Np_s)^2/Np_s \tag{11-16}$$

Note that the denominator of each term of the sum is just the expected number in that category and that the numerator is the square of the difference between the observed and expected distributions. Computationally, this can be simplified by noting that

$$\sum_{i=1}^{k} p_i = 1 \tag{11-17}$$

$$\sum_{i=1}^{k} Y_i = N \tag{11-18}$$

which by expansion of Eq. (11-16) and rearrangement yields

$$\chi^2 = (1/N) \sum_{i=1}^{k} [(Y_i/P_i)^2] - N \tag{11-19}$$

Having computed χ^2, one determines whether it is reasonable or not by consulting a χ^2 distribution table giving the chi-square distribution with v degrees of freedom. The number of degrees of freedom is $k - 1$ rather than the number of categories k. The choice of $k - 1$ rather than k can be justified intuitively. Not all values of Y are independent, since if $k - 1$ of them are known, the remaining one can be computed. Table 11-2 is an abbreviated χ^2 distribution table. More extensive tables are given by Knuth and by Bevington. The table should be read as follows. If one reads a value for row v and column p, then there is a probability p that χ^2 will assume a greater value than the entry. Thus, for example, in one test of a uniform random number generator with $k = 10$, χ^2 was found to be 8.30. Thus, from the table for $v = 9$, the probability that χ^2 will assume this value is very nearly 50%, which is acceptable. On the other hand, for another generator we found $\chi^2 = 48.22$. The probability of finding a χ^2 greater than 27.9 is $<0.1\%$. We thus conclude that this generator is almost certainly unacceptable. Perhaps somewhat surprisingly, a χ^2 of 1 would be equally suspect. This is true since $>99\%$ of the time χ^2 should be larger than this value. In other words the conformity to the expected value is too good.

TABLE 11-2

Selected Values of the Chi-Square Distribution[a]

v \ $p(\%)$	99	95	75	50	25	5	1	0.1
1	0.00016	0.00393	0.1015	0.455	1.323	3.841	6.635	10.8
3	0.115	0.352	1.21	2.37	4.11	7.82	11.3	16.1
8	1.65	2.73	5.07	7.34	10.2	15.5	20.1	26.1
9	2.09	3.33	5.90	8.34	11.4	16.9	21.7	27.9
15	5.23	7.26	11.0	14.3	18.3	25.0	30.6	37.7
35	18.5	22.5	29.0	34.3	40.3	49.8	57.2	66.6
48	28.2	33.1	41.0	47.3	54.3	65.2	73.7	84.0
100	70.1	77.9	90.0	99.3	109.3	124.3	135.8	149.4
200	156.4	168.2	186.0	199.4	213.3	224.0	249.4	267.6

[a] All results are derived from Bevington (1969, Table C-4). Copyright 1969 by McGraw-Hill.

In practice, how would the χ^2 test work? N should be relatively large. Knuth (1969) has recommended $Np_s > 5$. This condition is necessary because χ^2 is a valid statistic only for large N's. In practice, values of N of 10^3 or greater are desirable. Caution must be exercised, however, not to make N too large, because too large an N can smooth out local nonrandomness in the overall global uniformity of the generator. Since it is the local nonrandomness that usually damages simulation, it is best to make a number of runs with a relatively small N and verify that the distribution in the χ^2 does not deviate greatly from the table predictions. Further, even if the first few χ^2's look reasonable, this is no guarantee that a particular generator is adequate; different portions of the pseudorandom number sequences may exhibit a strong pattern.

Finally, if the random number generator utilizes some piece of hardware, such as a disk drive, for generating the random number, it is particularly important to repeat the test many times. This is to ensure that all hardware states are free of problems. For example, one of the author's systems has a random number generator that interrogates the disk status. If the disk is not used between calls, the generator frequently generates unacceptable sequences. Perhaps 5% of the time, however, perfectly acceptable sequences arise. Thus, it would have been entirely possible to have turned the system on and concluded that this generator was acceptable when in fact it has turned out to be the most unacceptable generator we have encountered.

3. The Kolmogorov–Smirnov Test

The χ^2 test weighs all the deviations between the observed and expected distributions. Therefore, it has a tendency to overlook local defects in the distribution because a local region of very poor fit can be compensated for by regions of too good a fit. The KS test examines the maximum deviations between the expected and observed distribution functions; it is thus sometimes more sensitive for detecting local defects. The distribution function $F(x)$ of a random quantity X is given by

$$F(x) = \text{probability that } X < x \tag{11-20}$$

For example, the uniform random distribution is given by

$$F(x) = \begin{cases} x, & 0 \le x \le 1 \\ 0, & x < 0 \\ 1, & x > 1 \end{cases} \tag{11-21}$$

For a collection of N data points $X_1, X_2, \ldots, X_N$, the distribution function is given by

$$F_N(x) = (\text{number of } X_i\text{'s} < x)/N \qquad (11\text{-}22)$$

To make the test we evaluate K_N^+ and K_N^- from

$$K_N^+ = N^{1/2} \max[F_N(x) - F(x)] \qquad (11\text{-}23a)$$

$$K_N^- = N^{1/2} \max[F(x) - F_N(x)] \qquad (11\text{-}23b)$$

where max denotes the maximum deviation for the term over the entire distribution, K_N^+ is related to the maximum observed positive deviation from the expected $F(x)$, and K_N^- is related to the maximum negative deviation. As with χ^2 there is an expected distribution in K_N^+ and K_N^-. Table 11-3 summarizes several values. A more complete table is given in Knuth. As in Table 11-2, for a given entry under p, one reads that there is p probability that K_N^+ or K_N^- will be larger than the table value.

We outline a detailed test for evaluating K_N^+ and K_N^-. Collect N values of the variable X. Order them in ascending values such that $X_1 < X_2 < \cdots < X_N$. Evaluate K_N^+ and K_N^- from

$$K_N^+ = N^{1/2} \max[(j/N) - F(X_j)], \qquad j = 1, \ldots, N \quad (11\text{-}24a)$$

$$K_N^- = N^{1/2} \max\{[F(X_j) - [(j - 1)/N]\}, \qquad j = 1, \ldots, N \quad (11\text{-}24b)$$

Again as with the χ^2 test a number of experiments should be run to verify that there are no regions or times after initialization of the generator when it fails badly. Further, to characterize the generator exhaustively, a number of runs should be made and the distribution of K_N^+'s and K_N^-'s should conform to the distribution predicted in Table 11-3.

TABLE 11-3

Selected Values of the K_N^+ and K_N^- Distribution[a]

N	$p(\%)$ 99	95	75	50	25	5	1
50	0.050	0.140	0.36	0.57	0.81	1.20	1.49
100	0.056	0.146	0.36	0.57	0.82	1.21	1.50
200	0.060	0.150	0.37	0.58	0.82	1.21	1.50
1000	0.066	0.156	0.37	0.58	0.83	1.22	1.51

[a] Data were calculated from the approximate expression given by Knuth (1969).

4. Sequence Tests

Distribution tests only verify that over relatively long sequences the average distribution matches relatively closely the expected distribution. No test of the order of occurrence of the runs was made. For example, if a coin were tossed 12 times and one obtained the distribution HTHTHTHTHTHT or HHTTHHTTHHTT, one would immediately be suspicious of the great regularity even though the expected distribution of heads and tails is correct. The three sequence tests described below are the serial test, the ordinary runs test, and the runs up/down test.

5. Serial Test

The serial test, which is also a distribution test, searches for correlations between two successive random numbers. It is designed to test sequences of integers where each integer has equal probability of occurring. The series is Y_1, Y_2, Y_3, Y_4, . . . , Y_{2N-1}, Y_{2N}, where the Y's are integers spanning the range 1 to d. Succeeding pairs of Y's, (Y_1, Y_2), (Y_3, Y_4), (Y_5, Y_6), . . . , (Y_{2N-1}, Y_{2N}), should be unrelated and therefore all possible pairs (q, r) with q and r running from 1 to d should be equally likely. For example, if $d = 2$, one would expect an equal probability of finding the pairs $(1, 1)$, $(1, 2)$, $(2, 1)$, and $(2, 2)$.

To carry out the serial test we count the number of times the pair (Y_{2j-1}, Y_{2j}) equals (q, r) for $j = 1$ to N. If the integers span the range 1 to d, this is exceptionally easily done on a computer as follows. Set up a $d \times d$ two-dimensional array $\mathbf{A}(d, d)$, zero all elements, and then do the counting:

$$\mathbf{A}(Y_{2j-1}, Y_{2j}) = \mathbf{A}(Y_{2j-1}, Y_{2j}) + 1, \qquad j = 1, . . . , N \quad (11\text{-}25)$$

You then have a $d \times d$ matrix with each element $\mathbf{A}(q, r)$ equal to the number of times the pair (q, r) occurred. All $\mathbf{A}(q, r)$'s should be equally probable, and we can carry out a χ^2 test with $v = d^2 - 1$ degrees of freedom with an expected probability for each element of $1/d^2$. For reasonable statistics N should be greater than $5d^2$; we have found $N = 20d^2$ to be a reasonable compromise in terms of computational speed and reliable statistics. It is important to note that each pair must be independent of every other pair. Therefore, we cannot use the following pairs: (Y_1, Y_2), (Y_2, Y_3), (Y_3, Y_4).

We found the serial test particularly valuable in evaluating hardware random number generators based on reading the current contents of a very high-speed counter. If successive readings are taken at intervals that are long with respect to the repeat cycle of the counter, such a scheme will yield an exceptionally good random number sequence. If, however, the read cycles are too close together, the clock count will have incremented only a small number on the average. Strong serial correlations then occur, since each succeeding element is correlated with its predecessor. The serial test is exceptionally good at detecting the correlation even when the random numbers themselves appear visually excellent.

As before, where an exceptionally good generator is required, it is not adequate to run the serial test a few times and conclude from the absence of values with improbable χ^2's that the generator is good. A much more satisfactory but quite time-consuming approach is to calculate a large number of χ^2's and compare the observed distribution with the expected one. How many χ^2's should be evaluated? The answer depends on how bad the generator is. For example, a marginal generator shows clear inadequacy in the distribution of χ^2's after 5–10 χ^2's. A nearly acceptable generator might require 100–500 χ^2's for its inadequacy to become apparent. From 1000 to 10,000 trials might be required to be certain of a generator.

6. Runs Tests

Runs tests look for repeating patterns. Unlike the serial test, however, runs tests look at the continuation over long distances rather than in just adjacent elements. For example, ten heads on ten consecutive coin tosses would be very suspicious. We discuss both the ordinary runs test and the runs up/down test. The ordinary runs test requires a variable which can be treated as having only two states (e.g., 0, 1; heads, tails; positive, negative) both having equal probability. The runs up/down test treats continuous distributions. In both cases the observed number of runs is compared with the expected ones and both can find nonrandomness missed by the distribution or serial test.

a. Ordinary Runs Test

If there are two equally probable states of a random number generator, a run is defined as an unbroken sequence of a single state or character of the system. A single isolated character in the string is a run of 1. For example the sequence 10110111000 has a total of six runs, with three runs of 1's and three runs of 0's.

We will worry here only about long sequences with >50 values, since for shorter sequences the treatment is more complex and we have found that the range in acceptable runs is so large in general as to make the test useless. We let V be the total number of elements of one type and T the total number of elements of the other type in the sequence. N is the total number of values tested and equals $V + T$. R is the total number of runs. For long sequences the expected distribution in R will approach a normal distribution with an expected number of runs R_{exp} and a standard deviation in the number of runs σ_{run} given by (Gibbons, 1976)

$$R_{exp} = 1 + 2VT/N \qquad (11\text{-}26a)$$

$$\sigma_{run} = [2VT(2VT - N)]/[N^2(N - 1)] \qquad (11\text{-}26b)$$

Thus, for a normal distribution there is a 50% probability that a single observed R will fall within $\pm 0.6745\sigma_{run}$ of R_{exp}. Similarly, the probabilities of falling in the following ranges are $R_{exp} \pm \sigma_{run}$ (68.3%), $R_{exp} \pm 1.96\sigma_{run}$ (95%), $R_{exp} \pm 2.58\sigma_{run}$ (99%), and $R_{exp} \pm 3.3\sigma_{run}$ (99.9%).

Thus, the ordinary runs test is performed as follows: Choose a relatively large N, determine the number of runs of each of the two types V and T, calculate the total number of runs $R = (V + T)$, and compare these with the R_{exp}. If, after repeating the test a number of times, the distribution of R agrees reasonably closely with the above-listed distribution, then the generator passes the test. Frequently omitted in discussions of the runs test is that V and T should be about equal—that is, the distribution should pass a distribution test. The author has seen a very poor random number generator that routinely gave $V \sim 400$ and $T \sim 600$, yet the distribution in R was always fully acceptable. Thus, the runs test, especially the ordinary runs test, should never be used as the sole basis for acceptance of a generator. At the very least, it must be used with a distribution test.

Statistically, the ordinary runs test is considered a weak test because it uses so little detail about the distribution. However, it is an exceptionally powerful test for testing individual bits in a binary word which are, by nature, completely defined as two states. Each individual bit of the final binary word should pass the runs test.

We conclude with a brief discussion of how to convert a continuous distribution into a two-state one with equal probability of either occurrence. X, which is continuously and uniformly distributed over the interval $0 \le X \le 1$, is easily converted by making the variable in the runs test 0 if $0 \le X < 0.5$ and 1 if $0.5 \le X < 1$. Another example is if the original distribution D is of the integers 0–9, each with the same probability of occurrence. A natural division is to assign a 0 if D is less than the median of 4.5 and 1 if it is greater. The division need not be so obvious, however. If one believed that there was a tendency for 0, 7, and 9 to cluster and for

1, 2, and 8 to cluster, then a more definitive test would be to assign a 0 if $D = 0, 3, 4, 7,$ or 9 and a 1 if $D = 1, 2, 5, 6,$ or 8. A normal distribution with a mean of 0 and a standard deviation of 1 can be checked by counting the number of elements falling in the range -0.6754 to $+0.6754$ and 1 to those falling outside this range.

b. Runs Up/Down Test

The ordinary runs test forces elements in a sequence to assume only two values. This restriction can exclude information contained in the relative magnitude of elements. In the runs up/down test, the magnitude of each element is compared with the previous element in the sequence. If the preceding element is smaller, we have a run up and if the preceding element is larger, we have a run down. For example, the sequence 1, 2, 3, 4, 2, 4 has two runs up; one is of length 3 (1, 2, 3, 4) and the other of length 1 (2, 4). It also has a single run down of length 1 (4, 2). A convenient way of counting runs up and down is to generate a new sequence in which the element 1 is generated for two adjacent elements running up and a 0 for every pair of elements running down. Thus, the previous sequence becomes 1, 1, 1, 0, 1. This simplified sequence is one number shorter than the original, and the total number of runs up equals the number of contiguous groups of 1's and the number of runs down equals the number of groups of contiguous 0's.

The test is carried out in the following way. Count the total number of runs either up and down to obtain R. Now for large values of N (>25) the expected distribution of R is normal with a mean value R_{exp} and a standard deviation σ_{exp} given by (Gibbons, 1976)

$$R_{exp} = (2N - 1)/3 \tag{11-27a}$$

$$\sigma_{exp} = [(16N - 29)/90]^{1/2} \tag{11-27b}$$

As with the ordinary runs test one chooses a large value of N (typically 500–5000 to make σ_{exp} small enough to be useful), calculates R, and compares its deviation from R_{exp} in units of σ_{exp}. Thus, about 50% of the trials should fall within $R_{exp} \pm 0.6745\sigma_{exp}$. Only ~0.1% of the R's should fall outside $R_{exp} \pm 3.3\sigma_{exp}$.

Note that, unlike the ordinary runs test, R_{exp} is dependent only on N and independent of the number of elements of different types. As with the ordinary runs test, a generator should be considered unacceptable if it is either too regular (too few runs) or too random (too many runs), since both indicate regularity that does not exist, on the average, in real random number sequences.

7. A Test of Random Number Generators

To demonstrate the application of these statistical tests, we have applied all the tests described here to the random number generators used to generate the two "uniform" distributions shown in Fig. 11-2. Generator 1 (Fig. 11-2a) is the random number generator of a Hewlett-Packard 9825A desk-top calculator. To date it is the best generator we have examined. It has successfully passed most of the statistical tests we have applied to it, including ones far more exhaustive than those described here. Generator 2 (Fig. 11-2b) is a deliberately distorted one with multiple defects built into it. We have, however, seen random number generators on other systems that are equally bad.

In the interest of reasonable computational times, each statistical test was performed only ten times. Unless otherwise indicated, sequences of 1000 random numbers were tested. Rather than reporting means for each test we report medians. For such small samples, the median is less sensitive than the mean to the presence of statistically possible outliers, and, further, the median can be compared directly with expected values for $p = 50\%$.

We began by evaluating the mean and standard deviation for each generator (Table 11-4). For the good generator 1, the median of the ten means was 0.5015 with values ranging from 0.4913 to 0.5181. Since the expected standard deviation of the mean is 0.0091 ($0.2886/1000^{1/2}$), the median mean and spread in means of generator 1 clearly falls within the expected range. The bad generator 2 yields a median mean of 0.4922 and a range of values for 0.4721–0.5031. These values are suspicious. The lowest observed mean is too low by 3.1 standard deviations of the mean, and there appears to be a shortage of values above 0.5000.

The standard deviations are also revealing. The good generator 1 has a spread of 0.2799–0.2914 with a median of 0.2887, which agrees very well with the expected 0.2886. On the other hand generator 2 has somewhat too high a median standard deviation and all values are above the expected ones.

At this stage of the testing there are strong grounds to suspect generator 2, but because of the small sampling and the relatively small discrepancies it is still not clear that generator 2 has failed. Further tests of generator 2 are clearly required. Ideally, longer runs should be tested to reduce the expected standard deviations of the mean to make the tests more definitive. Alternatively, other tests could be used. Further tests may be misleading, however, as it is potentially possible to fail the mean and standard deviation tests but pass others.

TABLE 11-4

Results of Tests on Two Random Number Generators

GENERATOR#	Mean & Standard Deviation [a]				Bin Sort [b]		Serial Test [c]		KS Test [d]				Ordinary Runs [e] Test		Runs Up/Down [e] Test	
	1		2		1	2	1	2	1		2		1	2	1	2
TRIAL	μ	σ	μ	σ	χ^2	χ^2		χ^2	K^+	K^-	K^+	K^-	ND		ND	
1	0.4925	0.2871	0.4721	0.3041	5.1	48.2	18.5	31.5	0.20	1.22	1.11	1.65	-0.72	2.16	-0.38	2.18
2	0.5150	0.2908	0.5032	0.3016	8.3	34.4	18.4	54.6	0.26	1.68	1.65	0.92	0.75	2.86	-0.90	1.73
3	0.4913	0.2914	0.4867	0.3051	9.2	33.4	12.8	42.3	0.48	0.51	1.47	1.44	0.38	1.71	-0.53	-0.15
4	0.5107	0.2353	0.4982	0.3041	14.2	31.8	8.7	27.2	0.83	0.71	2.04	0.54	2.06	2.36	-0.15	-0.53
5	0.4991	0.2910	0.4982	0.3000	5.8	44.4	20.6	34.8	0.75	0.83	2.05	0.62	0.89	1.51	0.53	1.35
6	0.4996	0.2900	0.4787	0.3018	7.2	43.5	16.7	35.9	0.75	0.57	1.51	1.42	0.45	1.89	0.53	0.38
7	0.5160	0.2885	0.4928	0.3074	9.2	41.9	11.8	26.1	0.98	0.49	0.83	1.62	-0.05	2.72	1.13	0.68
8	0.5096	0.2390	0.4958	0.2999	2.3	37.3	7.2	67.0	0.26	0.91	1.47	1.46	0.34	1.74	-0.15	-1.43
9	0.5181	0.2799	0.5018	0.3025	5.9	36.8	19.4	27.6	0.62	0.27	1.26	1.33	-0.43	3.82	-0.83	1.05
10	0.5031	0.2827	0.4915	0.2990	5.1	43.7	12.2	39.5	0.49	0.74	1.96	0.83	1.53	3.09	-0.98	2.00
Median(obs)	0.5048	0.2887	0.4943	0.3033	6.5	39.6	14.8	35.3	0.69	0.72	1.49	1.38	0.36	2.26	-0.27	0.87
Median(exp)	0.5000	0.2886	0.5000	0.2886	8.3	8.3	14.3	14.3	0.59	0.59	0.59	0.59	0.00	0.00	0.00	0.00

[a] Each entry is the result of the evaluation of 1000 random numbers.

[b] Ten intervals were used.

[c] Four equal sized intervals were used. Each entry is calculated for 200 pairs.

[d] $N = 1000$

[e] 1000 numbers were used in each sequence.

We next carried out a bin sort with a chi-square test. Table 11-4 summarizes the results of a bin sort test for each generator using 10 evenly spaced bins (9 degrees of freedom). Generator 1 passes well with respect to both the median χ^2 and the distribution. Generator 2 fails abysmally. The lowest observed χ^2 falls well outside the $p = 1\%$ level. Thus, generator 2 clearly fails as a uniform random number generator. It is, however, significant to note that generator 2 easily passes a bin sort using two evenly spaced intervals. Thus, it is important when using a bin sort test to use at least two values of bins.

Table 11-4 also summarizes the results of a serial test using four integers ($d = 4$); 200 pairs were used for each test. Generator 1 passes as before, but generator 2 fails with the median χ^2 falling above the level expected for $p = 1\%$. Note, however, that two of the χ^2's for generator 2 fall in the $p = 5$ to 1% range. Thus, if these values were the first to appear, the incautious might assume that they were just outliers of an acceptable distribution and stop testing. In reality, they are outliers of a bad distribution. Thus, the need for making more than one or two tests of each type is strongly emphasized.

The results for a KS test with $N = 1000$ are also shown in Table 11-4. Generator 1 passes both the K^+ and K^- tests. Generator 2 clearly fails both in terms of the median (in the $p = 5$ to 1% range) and distribution. In particular, however, even for generator 2 there were a significant number of acceptable K^+'s and K^-'s, although never in the same run. Figure 11-3 is a plot of typical distribution functions ($N = 1000$) for generators 1 and 2.

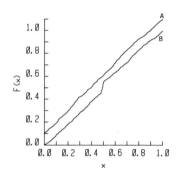

Fig. 11-3. Integral probability distribution functions A and B for good [(a) generator 1] and poor [(b) generator 2] uniformly distributed random number generators of Fig. 11-1. The data for generator 1 have been displaced by 0.1 to aid viewing. The theoretical distribution in both cases is given by a straight line $F(x) = x$ ($0 \le x \le 1$). Each distribution was derived from 1000 random numbers. For the plotted distributions, the K's are as follows: $K^+ = 0.367$ and $K^- = 1.124$ (generator 1); $K^+ = 0.923$ and $K^- = 1.681$ (generator 2).

Generator 1 shows no obvious defects and closely follows the expected $F(x) = x$ $(0 \leq x \leq 1)$ line. The defect in generator 2 in the region 0.45–0.55 is clearly evident in this case.

To test the sensitivity of the KS test to the value of N, we repeated the tests using $N = 50$ and $N = 200$. For $N = 50$ and only ten tests, both generators 1 and 2 appeared acceptable with respect to median and distribution. For $N = 200$, generator 1 was satisfactory and there was a suggestion that the median of generator 2 was too high. On the basis of just ten experiments, however, it would not have been possible to reject generator 2. Thus, in the current case, the KS test was not a strong test. Either N must be large to obtain good statistics, or if N is small a large number of replicates must be performed to give a good distribution in the K^+ and K^- tests.

For presenting the runs tests, we calculate the expected number of runs R_{exp} and the standard deviation σ_{exp} in the expected number of runs. We then display normalized data of the form

$$ND = (R - R_{exp})/\sigma_{exp}$$

This function should be approximately Gaussian with a mean of 0 and a standard deviation of 1. Thus, we would expect the median to be near 0 with ~95% of the results falling within ±2 and ~99.7% of the results falling within ±3. Generator 1 appears reasonable in the ordinary runs test, although the number of runs may be slightly excessive. Generator 2 clearly fails. There are always too many runs, and 60%, rather than the expected 5%, of the results fall outside ±2. Thus, generator 2 oscillates too often around 0.5000; it is too noisy to be random.

Interestingly, in the runs up/down test, generator 2 comes close to passing. The median is a little high, but the number of outliers is only marginally suspicious (i.e., 20% outside the $p = 5\%$ range of ±2). Generator 1 has a more reasonable median, but in fact has one value at 2.48. Thus, at this stage one might be somewhat suspicious of both generators.

It is particularly significant that generator 2 fails the ordinary runs test so badly, but perhaps only marginally fails the runs up/down test. It is clear that these two runs tests evaluate quite different properties of the random number generators, and the passing of one test alone is certainly no guarantee of an acceptable generator.

We will conclude this section by describing the results of a more thorough serial test on the good generator 1. Generator 2 is so poor that it is clear that more extensive testing is not warranted. What we have done is to carry out 10,000 serial tests using four equally probable categories $(d = 4)$. One expects the resultant distribution in χ^2's to be a probability distribution function of χ^2 with 15 degrees of freedom $(4^2 - 1)$. We then compare the resultant distribution with the expected one [Bevington (1969, Eq. (10-4))].

Figure 11-4 shows the observed and expected distributions. Visually the agreement appears quite acceptable. Further, a comparison of the integral probability distribution of the data agrees well with the theoretical distribution. For example, from Table 11-2 one expects 1% of the χ^2 to fall below $\chi^2 = 5.2$, and the observed value is 1.17% below 5.5. Twenty-five percent of the χ^2 should fall below $\chi^2 = 11.0$, and 26.4% are found below 11.5. Half of the χ^2's are expected for $\chi^2 < 14.34$, and 48.9% are found below $\chi^2 = 14.5$. Seventy-five percent of the χ^2 should occur below $\chi^2 = 18.25$, and 73.7% were found below $\chi^2 = 18.5$. Ninety-nine percent of the χ^2 should have occurred below $\chi^2 = 30.58$, and 99.0% of them were found for $\chi^2 < 32.5$. Finally only 0.1% of the χ^2's were expected to occur above $\chi^2 = 37.7$ and, in fact, only 0.15% of them were above 37.5. Thus, it appears that generator 1 is fully acceptable in this test.

Closer analysis of the data of Fig. 11-4, however, reveals that they are probably flawed. The observed distribution of χ^2's should also pass a χ^2 distribution test itself. On carrying this out we obtain a $\chi^2 = 183$ for 48 degrees of freedom. The expected value for $p = 0.1\%$ is 84. Thus, generator 1 is in fact not a perfect generator. We also originally carried the test out on 1000 χ^2s. The distribution was similar, but of course noisier than that of Fig. 11-4. Interestingly, the lower signal-to-noise ratio masked the defect and gave an acceptable χ^2 test for the χ^2 distribution.

We have also applied a χ^2 test to the Gaussian noise generated in Fig. 11-2. This data set gives much too large a $\chi^2 \sim (450)$ versus the expected 146 for $p = 0.1\%$. Thus, the Gaussian distribution derived from the some-

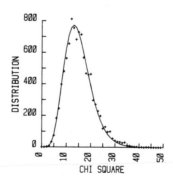

Fig. 11-4. Distribution in χ^2's for 10,000 serial tests on random number generator 1. The +'s are histogram plots of the number of χ^2's falling in an interval 1 unit wide and centered on every integer value. The solid line is the theoretical χ^2 curve (Bevington, 1969). A 4 × 4 array (15 degrees of freedom) with 200 pairs of random numbers was used for each test.

what flawed Hewlett-Packard 9825A uniform random number generator is clearly also imperfect.

We have two observations from these results. First, virtually any pseudorandom number generator and probably most hardware random number generators will fail statistical tests for randomness if one looks hard enough and long enough. The calculations required to generate the data of Figs. 11-2 and 11-4 each took ~8 h on the 9825A calculator, which is three to ten times faster than most 8-bit microcomputers. Second, just because a generator fails a test does not automatically exclude its use. Examination of the data of Fig. 11-4 shows that it fails because of the irregularity of the χ^2 distributions rather than because of a large number of fits that are either too good (low χ^2's) or too bad (high χ^2's). Similarly, the Gaussian noise of Fig. 11-2 fails not because of outliers but because of excessive noise on the distribution functions. In these cases we would not expect these failures of generator 1 to affect its use significantly in simulations. On the other hand, had a significant number of outliers resulted (e.g., low or high χ^2's in Fig. 11-4), we would have had valid reasons for concern about the use of this generator in simulations.

8. Summary of Random Number Tests

No single test should form the basis of acceptance of a random number generator except for the crudest calculations or games. Any generator must, at the very least, give the correct mean and standard deviation and pass a χ^2 distribution test with several different numbers of bins. We strongly recommend at least a few runs with the serial test and both the ordinary runs test and the runs up/down test. For the linear congruential generators common to many computer software generators, Knuth (1969) has pointed out that most will pass the χ^2 and serial correlation tests. Bad generators of this type are much more frequently ferreted out by the runs up/down test. With improper selection of parameters in the generator, the runs tend to be excessively long. Conversely for hardware generators based on high-speed counters, we find the serial test to be especially useful, although here the runs tests are very powerful.

Even after satisfactorily passing all the tests, a random number generator must still be regarded with suspicion. Only after it has been used extensively and no inexplicable problems in its use have arisen should it be considered a safe generator.

The way in which the noise generator is called should also be taken into account when testing the generator. If the random number generator is called in batches, the test should be done somewhat differently. For example, two random numbers X_j and X_{j+1} may be required at a time. This is true if noise is to be added to $E(t_i)$ and $D(t_i)$. In this case only alternate calls to the generator would be used to add noise to $E(t_i)$. Thus, in addition to making tests on every element in the sequence, it would be prudent to make the standard tests on the two subsequences X_1, X_3, X_5, . . . and X_2, X_4, X_6, ... This procedure ensures that alternate term periodicity or other nonrandomness that is masked by examining every term does not go unnoticed and introduce systematic bias on the individual curves.

E. CHEMICAL STANDARDS

Digital simulations are ideal for testing a new mathematical data reduction scheme or for training an operator in the use of a method. They do not, however, test overall instrument performance. It is, therefore, desirable to have at least a few chemical standards in order to test a system. The ideal standard should be available in high purity, be inexpensive, be convenient to use, and have a well-known lifetime. Ideally, it should also have a high quantum yield to minimize purity problems. A recent comprehensive list and verification of lifetime standards has been published (O'Connor et al., 1983).

While there seems to be no truly trustworthy standard in the subnanosecond range, the proliferation of good single-photon counting instruments has supplied a wealth of molecules suitable as nanosecond or longer standards. Ware (1971) and Berlman (1971) have listed a number of molecules, as have Rayner et al. (1976).

Quinine in 0.1–1.0 N sulfuric acid has been used as a relatively long-lived standard. Typical τ's are 19.5 nsec (TRW Systems, 1967), 19.7 nsec (average of two literature values; Ware, 1971), 19.3 nsec (Wild et al., 1977; Isenberg et al., 1973), and 19.4 nsec (Love and Upton, 1980). O'Connor et al. (1982), using a subnanosecond-response system, have found that quinine in 0.1 N or 1.0 N sulfuric acid has a wavelength-dependent double-exponential decay. The long-lived component varies somewhat with wavelength and acid concentration but is ~19–20.7 nsec. Single-exponential fits near the emission maxima (450 nm) yielded 19.7 nsec (1.0 N acid) and 19.1 (0.1 N acid). Thus, as an ~20 nsec reference, quinine is a reasonable material, but it should not be treated as a precision reference.

For short lifetimes the scintillator POPOP appears to be an excellent standard in deoxygenated cyclohexane. Reported lifetimes are 1.10–1.15 nsec (O'Connor et al., 1979; Andre et al., 1979) and 1.10 nsec (Rayner et *al.*, 1976). O'Connor *et al.* (1983) have recommended PPO with a lifetime of 1.42 nsec in deoxygenated cyclohexane.

Another molecule that should be an excellent standard is anthracene. In deoxygenated ethanol the values are reasonable and include 5.5 nsec (Ware, 1971–average of four literature values), 5.67 nsec (Wild *et al.*, 1977), and 4.93 nsec (Rayner *et al.*, 1976). In dilute solutions in deoxygenated cyclohexane, however, values include 3.97 nsec (Hazan *et al.*, 1974), 5.3 nsec (O'Connor *et al.*, 1979), 5.23 nsec (O'Connor *et al.*, 1983), and 5.03 nsec (Rayner *et al.*, 1976), which deviate more than one would like. In ethanol a value of 5.3 nsec and in cyclohexane a value of 5.2 nsec seem appropriate. The wide spread of results, especially in cyclohexane, should be kept in mind.

Standards are also useful in testing single-photon counting instruments even if the lifetime is not well known. If the reference is known to have a well-behaved exponential impulse response, it still makes a good test. If the instrument is behaving well, the sample decay should be fit well over several decades. Failure to obtain such a fit suggests an improperly working instrument. It is not necessary to know the reference lifetime accurately to apply this test. All of the above-mentioned standards are useful in this type of test.

A

Solution of the Generalized Response [Eq. (4-2)]

Equation (4-1) is a first-order ordinary differential equation of the form

$$dy/dt + p(t)y = q(t) \tag{A-1}$$

It has an exact solution that can be solved by the integrating factor method (Kaplan, 1952):

$$s(t)y = \int q(x)s(x) \, dx + c,$$
$$s(t) = \exp[\int p(t) \, dt] \tag{A-2}$$

where y is [*D], $p(t)$ is the constant k, and $q(x)$ is $I(t)$.

$$[*D] = e^{-kt}\{\int [\exp(kx)]I(t) \, dx + c\} \tag{A-3}$$

where x is a dummy variable of integration and c is the constant of integration which is adjusted to satisfy the necessary boundary value conditions. By assuming [*D] is zero at $t = 0$, we obtain $c = 0$. Integrating from $x = 0$ to $x = t$ gives

$$[*D] = \exp(-kt) \int_{x=0}^{x=t} [\exp(kx)]I(x) \, dx \tag{A-4}$$

B

Solution of the Phase Shift Formula [Eq. (4-12)]

Substitution of Eq. (4-11a) into Eq. (4-3) with $K = 1$ yields

$$D(t) = B \exp(-kt) \int_0^t [\exp(kx)][1 + M \cos(\omega x)] \, dx = I_1 + I_2 \quad \text{(B-1a)}$$

$$I_1 = B \exp(-kt) \int_0^t \exp(kx) \, dx \quad \text{(B-1b)}$$

$$I_2 = BM \exp(-kt) \int_0^t [\exp(kx)] \cos(\omega x) \, dx \quad \text{(B-1c)}$$

I_1 is the transient solution for a step excitation:

$$I_1 = B[1 - \exp(-kt)] \quad \text{(B-2)}$$

The integral portion of I_2 is of the form (Pierce and Foster, 1957)

$$\int e^{ax} \cos(px) \, dx = \exp(ax)(a \cos px + p \sin px)/(a^2 + p^2) \quad \text{(B-3)}$$

$$I_2 = BM\{[\exp(kt)][k \cos \omega t + \omega \sin \omega t]\}/s \, \Big|_0^t$$

$$s^2 = k^2 + \omega^2 \quad \text{(B-4)}$$

$$I_2 = (BM/s)[(k/s) \cos \omega t + (\omega/s) \sin \omega t - (k/s) \exp(-kt)]$$

The two trigonometric terms in Eq. (B-4) can be simplified by use of the identity $\cos(\omega t - \delta) = \cos \delta \cos \omega t - \sin \delta \sin \omega t$. If we then make the assignment

$$\cos \delta = k/s, \qquad \sin \delta = \omega/s \qquad \text{(B-5)}$$

Then

$$\delta = \tan^{-1}(\omega/k), \qquad I_2 = (BM/s)[\cos(\omega t - \delta) - (k/s)\exp(-kt)]$$

Substitution of I_1 and I_2 into Eq. (B-1) yields Eq. (4-12).

C

Solution of Coupled Equilibria [Eq. (4-25)]

Our solution uses the approaches given in standard texts on differential equations (Kaplan, 1952). We first define

$$x = [*A], \qquad y = [*B]$$
$$x' = dx/dt, \qquad y' = dy/dt$$

X and Y are defined in Eqs. (4-26e) and (4-26f). Equations 4-24 now yield the system of linear equations

$$x' = -Xx + k_{BA}y \qquad \text{(C-1a)}$$

$$y' = k_{AB}x - Yy \qquad \text{(C-1b)}$$

In any system of the type that has constant coefficients, a particular trial solution is

$$x = D \exp(\lambda t) \qquad \text{(C-2a)}$$

$$y = E \exp(\lambda t) \qquad \text{(C-2b)}$$

where D, E, and λ are constants. Substitution of Eqs. (C-2) into Eqs. (C-1) and elimination of $\exp(\lambda t)$ yields the following pair of equations in D and E:

$$(-X - \lambda)D + k_{BA}E = 0 \qquad \text{(C-3a)}$$

$$k_{AB}D + (-Y - \lambda)E = 0 \tag{C-3b}$$

This is the set of characteristic equations associated with Eqs. (C-1). Besides the trivial solution $D = E = 0$, the only nontrivial solutions arise when the determinant of the coefficients of the equation is zero:

$$\begin{vmatrix} -X - \lambda & k_{BA} \\ k_{AB} & -Y - \lambda \end{vmatrix} = 0 \tag{C-4}$$

This yields the quadratic equation

$$(X + \lambda)(Y + \lambda) - k_{BA}k_{AB} = 0 \tag{C-5}$$

which has the solutions

$$\lambda_1 = 0.5\{-(X + Y) - [(X - Y)^2 + 4k_{BA}k_{AB}]^{1/2}\} \tag{C-6a}$$

$$\lambda_2 = 0.5\{-(X + Y) + [(X - Y)^2 + 4k_{BA}k_{AB}]^{1/2}\} \tag{C-6b}$$

From Eqs. (C-3a) we obtain two nontrivial solutions relating D and E:

$$D_1 = k_{BA}E_1/(X + \lambda_1) \quad \text{for } \lambda_1 \tag{C-7a}$$

$$D_2 = k_{BA}E_2/(X + \lambda_2) \quad \text{for } \lambda_2 \tag{C-7b}$$

Equations (C-7) yield only the ratio of D to E. We can, therefore, choose any nonzero E without loss of generality and set $E_1 = E_2 = 1$. Now, if Eqs. (C-7) are solutions, so are the following linear combinations of the particular solution.

$$x = c_1 D_1 \exp(\lambda_1 t) + c_2 D_2 \exp(\lambda_2 t) \tag{C-8a}$$

$$y = c_1 E_1 \exp(\lambda_1 t) + c_2 E_2 \exp(\lambda_2 t) \tag{C-8b}$$

where c_1 and c_2 are constants.

The constant coefficients c's in Eqs. (C-8) are solved for by use of the system's boundary value conditions. For an instantaneous flash, we have at $t = 0$

$$[*A]_0 = x_0 \tag{C-9a}$$

$$[*B]_0 = 0 \tag{C-9b}$$

Using Eqs. (C-8b) we immediately obtain

$$c_1 = -c_2 \tag{C-10}$$

which from Eqs. (C-8a) and (C-7) gives

$$x_0 = c_1 D_1 - c_1 D_2$$
$$= c_1 k_{BA}[(X + \lambda_1)^{-1} - (X + \lambda_2)^{-1}] \tag{C-11}$$

Equation (C-11) rearranges to

$$c_1 k_{BA} = x_0(X + \lambda_1)(X + \lambda_2)/(\lambda_2 - \lambda_1) \tag{C-12}$$

From Eqs. (C-8a) and (C-7), x is given by

$$x = \frac{c_1 k_{BA} \exp(\lambda_1 t)}{(X + \lambda_1)} - \frac{c_1 k_{BA} \exp(\lambda_2 t)}{(X + \lambda_2)} \tag{C-13}$$

which substituting for $c_1 k_{BA}$ from Eq. (C-11) yields

$$x = x_0[(X + \lambda_2) \exp(\lambda_1 t) - (X + \lambda_1) \exp(\lambda_2 t)]/(\lambda_2 - \lambda_1) \tag{C-14}$$

Since

$$(X + \lambda_1)(X + \lambda_2) = -k_{BA} k_{AB} \tag{C-15}$$

c_1 simplifies to

$$c_1 = -x_0 k_{AB}/(\lambda_2 - \lambda_1) \tag{C-16}$$

and

$$y = [x_0 k_{AB}/(\lambda_1 - \lambda_2)][\exp(\lambda_1 t) - \exp(\lambda_2 t)] \tag{C-17}$$

If we now set $\gamma_2 = -\lambda_2$ and $\gamma_1 = -\lambda_1$, we obtain Eq. (4-25).

APPENDIX

D

Phase Plane Equation with Scattered Light

With scattered light contributing to the observed decay, $D^{obs}(t)$

$$D^{obs}(t) = \alpha E(t) + D(t) \tag{D-1a}$$

$$D^{obs}(t) = \alpha E(t) + K \exp(-t/\tau) \int_0^t E(x) \exp(x/\tau)\, dx \tag{D-1b}$$

where $E(t)$ is the flash, $D(t)$ is the decay in the absence of scatter, and α is the scatter coefficient. The derivation follows that of Eq. (8-4). One takes the derivative of Eq. (D-1b) with respect to t and then integrates over the interval 0 to t, which yields

$$D^{obs}(t) = \alpha E(t) + K \int_0^t E(x)\, dx - (1/\tau) \int_0^t D(x)\, dx \tag{D-2}$$

Using Eq. (D-1a), Eq. (D-2) reduces to

$$Y(t) = A_0 + A_1 X_1(t) + A_2 X_2(t) \tag{D-3a}$$

$$A_0 = -1/\tau \tag{D-3b}$$

$$A_1 = K + \alpha/\tau \tag{D-3c}$$

$$A_2 = \alpha \tag{D-3d}$$

$$Y(t) = D^{obs}(t) \Big/ \int_0^t D^{obs}(x) \, dx \qquad \text{(D-3e)}$$

$$X_1(t) = \int_0^t E(x) \, dx \Big/ \int_0^t D^{obs}(x) \, dx \qquad \text{(D-3f)}$$

$$X_2(t) = E(t) \Big/ \int_0^t D^{obs}(x) \, dx \qquad \text{(D-3g)}$$

$Y(t)$, $X_1(t)$, and $X_2(t)$ are readily evaluated functions of $E(t)$ and $D^{obs}(t)$ using the trapezoidal rule method of Section 8-D. $Y(t)$ is linearly dependent on $X_1(t)$ and $X_2(t)$. The coefficients of Eq. (D-3a) yield the desired α, K, and τ. The normal phase plane equation [Eq. (8-4)] is just a special case of Eq. (D-3).

Data of the form of Eq. (D-3a) are readily fit by unweighted linear least squares. As in a normal linear least squares fit, one minimizes the function

$$\sum \{Y(t_i) - [A_0 + A_1 X_1(t_i) + A_2 X_2(t_i)]\}^2 \qquad \text{(D-4)}$$

One sets the three partial derivatives of Eq. (D-4) with respect to A_0, A_1, and A_2 equal to zero. This yields, in matrix form, the following normal equations

$$\begin{pmatrix} N & \sum X_1(t_i) & \sum X_2(t_i) \\ \sum X_1(t_i) & \sum X_1(t_i) & \sum X_1(t_i)X_2(t_i) \\ \sum X_2(t_i) & \sum X_1(t_i)X_2(t_i) & \sum X_2(t_i) \end{pmatrix} \begin{pmatrix} A_0 \\ A_1 \\ A_2 \end{pmatrix} = \begin{pmatrix} \sum Y(t_i) \\ \sum Y(t_i)X_1(t_i) \\ \sum Y(t_i)X_2(t_i) \end{pmatrix}$$

where the summations run over the N data points used in the fit. Solution of Eq. (D-5) for A_0, A_1, and A_2 coupled with Eqs. (D-3b) to (D-3d) yields α, K, and τ.

E

Computer Programs

Listings of three computer programs, EXPFIT, SIMPLEX, and LU-MEN, are given. The programs are written in BASIC for the HP85A desktop computer equipped with a matrix ROM and a printer/plotter ROM. The only unusual program aspects are the use of the SYS statement (which solves a system of linear equations in one line) and the graphics statements. Consult the HP-85 Owner's Manual and Programming Guide (00085-90002 Rev C7/80), the HP-85 Matrix ROM Manual (82937-9000) Rev. B 10/80), and the HP-85 Plotter/Printer ROM Manual (00085-90/40) for additional information. With the exception of LUMEN, which is used in the author's physical chemistry laboratory, the programs are written for demonstration purposes rather than as perfectly general programs. For example, in EXPFIT and SIMPLEX, fits are always carried out over the entire curves. A general program would permit fitting over a portion of the curve. Changes such as these are not difficult to implement.

EXPFIT is a simulation, nonlinear least squares fitting program. It can simulate decay data representing a single exponential or a sum of two exponentials exhibiting Poisson statistics. Weighted or unweighted fitting to one or two exponentials is done by the Marquardt method. One iteration requires about 10 sec with a single exponential and 45 sec with a double exponential.

SIMPLEX is a simulation and simplex minimization program using the algorithms given in the text. In SIMPLEX the program was adapted from the FORTRAN program given by Daniels.[†] The example involves fitting the excimer decay function $K[\exp(-t/\tau_1) - \exp(-t/\tau_2)]$ to a decay.

LUMEN uses the phase plane method to deconvolute a single exponential decay. It is based on the computer program described by Demas (1976) and by Wyker and Demas (1976). It was designed for manual entry of data and uses the Demas–Crosby (1970) convolution formula to regenerate the calculated decay curves.

[†] Adapted by permission of the publisher from Daniels (1978). Copyright 1978 by Elsevier North Holland, Inc.

```
01 ! ***********************************************************************
02 ! *                                                                    *
03 ! *                            EXPFIT                                   *
04 ! *                                                                    *
05 ! ***********************************************************************
06 !
07 ! BY J. N. DEMAS
08 ! LAST MODIFICATONS 1/19/83
09 ! SIMULATION AND REDUCTION OF SINGLE OR DOUBLE EXPONENTIAL DECAYS
10 ! DECAYS BY MARQUARDT NONLINEAR LEAST SQUARES FITTING
11 ! DECAYS CAN BE NOISE FREE OR EXHIBIT POISSON STATISTICS
12 ! DATA REDUCTION CAN BE EITHER WEIGHTED OR UNWEIGHTED
13 ! RESIDUAL PLOTS ARE SUPPLIED
14 !
15 ! ******************* MAIN VARIABLE USAGE **********************
16 ! N1=NUMBER OF EXPONENTIALS USED IN FIT (1 OR 2)
17 ! N=NUMBER OF VARIABLES USED IN FIT= 2*N1
18 ! N2= NUMBER OF EXPONENTIALS USED IN SYNTHETIC DATA
19 ! N4= FLAG UNWEIGHTED(1) WEIGHTED (2)
20 ! N9=MAX # OF ITERATIONS
21 ! E1=CONVERGENCE FRACTION IN A'S
22 ! E2=CONVERGENCE FRACTION IN CHI SQUARE
23 ! A()=UNDETERMINED PARAMETERS-CURRENT BEST GUESSES OF PARAMETERS
24 ! D1()=CORRECTION TO OLD GUESSES
25 ! T()=TIME ARRAY
26 ! D()=OBSERVED DECAY ARRAY
27 ! W()=ARRAY OF WEIGHTING FACTORS
28 ! R()=RESIDUALS TO SPEED EVALUATIONS
29 ! E(,)=ARRAYS OF EXPONENTIALS
30 ! S()=TEMPORARY STORAGE OF B MATRIX DIAGONAL ELEMENTS
31 ! S1()=TEMPORARY STORAGE OF OLD GUESSES IN CURRENT REFINEMENT
32 !        IN CASE CURRENT REFINEMENT FAILS TO IMPROVE CHI SQU
33 ! T=TIME INTERVAL BETWEEN POINTS
34 ! L=LAMBDA PARAMETER FOR MODIFICATION OF DIAGONAL ELEMENTS
35 ! B() C() LINEAR EQUATION ARRAYS
36 !     B()=MAIN ARRAY OF COEFFICIENTS
37 !     C()=COLUMN VECTOR ON RIGHT HAND SIDE OF EQUATIONS
38 ! E2=CHI SQUARE
39 !
40 ! PARAMETERS USED TO SIMULATE A DECAY
41 A1=10000 @ ! FIRST PREEXPONENTIAL FACTOR
42 A2=40 @ ! FIRST LIFETIME
43 A3=2000 @ ! SECOND PREEXPONENTIAL FACTOR
44 A4=7 @ ! SECOND LIFETIME
45 T=1 @ ! TIME INTERVAL BETWEEN POINTS
46 !
47 ! PARAMETERS USED TO SIMULATE A DECAY
48 ! COM IS AN HP QUIRK: MINIMIZES TAPE STORAGE
49 COM Z9
50 CLEAR
51 DISP "NONLINEAR LEAST SQUARES FIT ROUTINE BY MARQUARDT METHOD"
52 DISP "FITS 1 OR THE SUM OF 2 EXPONENTIALS TO EVENLY SPACED DATA"
53 DISP
54 DISP "THE FIRST POINT IS AT ZERO TIME"
55 N9=100 @ ! IF NOT CONVERGED AFTER 100 TIMES-TRY NEW GUESSES
56 L=.001 @ ! LAMBDA RECOMMENDED BY BEVINGTON (1969)
57 ! FRACTIONAL CHANGES IN PARAMETERS OR CHI SQUARE OF 1/10^5
58 ! SEEM MORE THAN ADEQUATELY TIGHT FOR GOOD CONVERGENCE
59 E1=.00001
60 E2=.00001
```

```
161 OPTION BASE 1
162 DIM A(4),D1(4),T(100),D(100),W(100)
163 DIM E(2,100),R(100),B(4,4),C(4),S(4),S1(4)
164 DISP "NUMBER OF COMPONENTS TO USE IN FITTING (1 OR 2)";
165 INPUT N1@ N=2*N1
166 !
167 ! ****************** GENERATE SYNTHETIC DATA ********************
168 GOSUB 363
169 !
170 ! REDIMENSION ARRAYS TO USE MATRIX ROUTINES
171 REDIM A(N),D1(N),B(N,N),C(N),S(N),S1(N)
172 !
173 ! **************** SET WEIGHTING FACTORS *********************
174 FOR J=1 TO 100 @ W(J)=1 @ NEXT J
175 DISP "WEIGHTING SCHEME:"
176 DISP "   1=UNWEIGHTED"
177 DISP "   2=SINGLE PHOTON COUNTING WEIGHT"
178 INPUT N4
179 IF N4=1 THEN 183
180 FOR J=1 TO 100 @ W(J)=1/D(J) @ NEXT J
181 !
182 ! SET ITERATION COUNTER
183 L9=0
184 !
185 ! ************* ENTER INITIAL GUESSES TO PARAMETERS ************
186 DISP "PREEXPONENTIAL FACTOR THEN LIFETIME FOR COMPONENT 1"
187 DISP "THEN FOR COMPONENT 2 IF NECESSARY"
188 FOR J=1 TO N @ DISP "A(";J;")";@ INPUT A(J)@ NEXT J
189 !
190 ! ****************** EVALUATE INITIAL CHI SQUARE **************
191 GOSUB 344
192 DISP "INITIAL CHI SQUARE= ";X2 @ DISP
193 X3=X2 @ ! SAVE LAST CHI SQU FOR COMPARISON
194 GOSUB 295 @ ! GENERATE  B() C() MAT COEFFICIENTS
195 FOR J=1 TO N
196 S(J)=B(J,J) @ ! SAVE OLD DIAGONALS OF COEFFICIENT MATRIX
197 S1(J)=A(J) @ ! SAVE OLD GUESSES
198 NEXT J
199 ! ADJUST DIAGONAL OF B() WITH LAMBDA THEN EVALUATE CORRECTIONS
200 FOR J=1 TO N
201 B(J,J)=B(J,J)*(1+L)
202 NEXT J
203 !
204 ! ************ SOLVE FOR CORRECTIONS TO A() GUESSES ***********
205 MAT D1=SYS(B,C)@ ! SOLVE SYSTEM OF LINEAR EQUATIONS
206 DISP "CORRECTIONS TO PARAMETERS"
207 FOR J=1 TO N
208 DISP "DELTA A(";J;")= ";D1(J)
209 NEXT J
210 !
211 ! ********** GENERATE CORRECTED A()'S AND DISPLAY THEM *********
212 MAT A=A-D1@ ! GENERATE CORRECTED PARAMETERS
213 DISP "TRIAL CORRECTED PARAMETERS"
214 FOR J=1 TO N
215 DISP "A(";J;")= ";A(J)
216 NEXT J
217 !
218 ! ***************** GENERATE TRIAL CHI SQUARE ****************
219 GOSUB 344 @ ! CHI SQUARE ROUTINE
220 DISP "CHI SQUARE";X2
```

```
221 ! IF CHI SQUARE ACTUALLY IMPROVED THEN REDUCE LAMBDA
222 ! CHECK FOR CONVERGENCE, MAKE TRIAL CHI SQUARE THE NEW ONE
223 IF X2<X3 THEN 240
224 ! IF YOU ARE HERE CHI SQUARE DIDN'T IMPROVE
225 ! RESTORE OLD B() DIAGONALS, INCREASE LAMBDA, TRY AGAIN
226 L=10*L
227 ! L MAY INCREASES WITHOUT BOUND- USUALLY ONLY FOR A GOOD SOLUTION
228 ! GIVE WARNING AND DISPLAYS FINAL RESULTS
229 IF L<=10000 THEN 231
230 DISP "WARNING LARGE L-CONVERGENCE MAY HAVE FAILED" @ GOTO 264
231 FOR J=1 TO N
232 B(J,J)=S(J)
233 A(J)=S1(J)
234 NEXT J
235 GOTO 200
236 !
237 ! ***********************************************************************
238 ! *                    TEST FOR CONVERGENCE                            *
239 ! ***********************************************************************
240 L=.1*L @ ! REDUCE L SINCE CHI SQUARE IMPROVED
241 F=0 @ ! SET FLAG FOR CONVERGENCE TEST 0=CONVERGED, 1=FAILED
242 ! CHECK TO SEE IF ALL A'S WITHIN SPECIFIED TOLERANCE
243 ! FLAG SET IF ANY OF THE DELTA A'S/A FAIL CONVERGENCE TEST
244 FOR J=1 TO N
245 IF ABS(D1(J)/A(J))>E1 THEN F=1
246 NEXT J
247 ! CHECK TO SEE IF CHI SQUARE CHANGE IN SPECIFIED TOLERANCE
248 IF ABS((X3-X2)/X2)<E2 THEN 255
249 IF F=0 AND ABS((X3-X2)/X2)<.05 THEN 255
250 X3=X2 @ ! OLD CHI SQU NOW NEW ONE
251 L9=L9+1 @ DISP "NO CONVERGENCE AFTER";L9;"ITERATIONS" @ DISP
252 ! IF DIDN'T CONVERGE AFTER LIMIT ON # ITERATIONS GO ERROR ROUTINE
253 IF L9>N9 THEN 285
254 GOTO 194
255 GOSUB 395 @ ! PLOT RESIDUALS
256 !
257 ! ***********************************************************************
258 ! *               CALCULATE ERRORS IN PARAMETERS                       *
259 ! ***********************************************************************
260 ! FIRST RESTORE B MATRIX TO ITS CORRECT FORM THEN INVERT IT
261 ! DIAGONAL ELEMENTS OF B INVERSE (ERROR MATRIX) ARE THE PARAMETER
262 ! VARIANCES FOR THE WEIGHTED CASE
263 ! FOR THE UNWEIGHTED CASE MUST BE CORRECTED (SEE TEXT)
264 FOR J=1 TO N @ B(J,J)=S(J) @ NEXT J @ ! RESTORE B()
265 MAT B=INV(B)@ ! GENERATE ERROR MATRIX
266 IF N4=2 THEN 278
267 ! UNWEIGHTED TREATMENT
268 Z1=0
269 FOR J=1 TO N
270 Z1=Z1+R(J)*R(J)
271 NEXT J
272 Z1=Z1/(100-N)
273 FOR J=1 TO N
274 DISP "ERROR A(";J;")";SQR(Z1*B(J,J))
275 NEXT J
276 END
277 ! WEIGHTED TREATMENT
278 FOR J=1 TO N
279 DISP "ERROR A(";J;")";SQR(B(J,J))
280 NEXT J
```

```
281 DISP "FINI HAVE A NICE DAY"
282 !
283 ! ***************** DIDN'T CONVERGE ***************************
284 END
285 DISP @ DISP "DIDN'T CONVERGE"
286 DISP "DO YOU WANT TO TRY NEW GUESSES (Y or N)";@ INPUT A$
287 IF A$="Y" THEN 183 ELSE DISP "BYE" @ END
288 !
289 ! *******************************************************************
290 ! *                 GENERATE MATRIX OF COEFFICIENTS               *
291 ! *******************************************************************
292 ! NOTE THAT ALL NECESSARY EXPONENTIALS ARE NOW STORED IN E()
293 ! ZERO MATRICES BEFORE GENERATING SUMS
294 ! Z1,Z2,Z3,Z4,Z5,Z6,Z7 ARE SCRATCHPADS
295 MAT B=ZER@ MAT C=ZER
296 FOR J=1 TO 100
297 Z1=W(J)*E(1,J)*E(1,J) @ B(1,1)=B(1,1)+Z1
298 B(1,2)=B(1,2)+T(J)*Z1
299 B(2,2)=B(2,2)+T(J)*T(J)*Z1
300 Z2=W(J)*E(1,J)*R(J) @ C(1)=C(1)+Z2
301 C(2)=C(2)+T(J)*Z2
302 NEXT J
303 Z3=A(1)/(A(2)*A(2))
304 B(1,2)=B(1,2)*Z3
305 B(2,2)=B(2,2)*Z3*Z3
306 C(2)=C(2)*Z3
307 B(2,1)=B(1,2)
308 IF N1=1 THEN RETURN
309 ! IF FITTING WITH TWO EXPONENTIALS GENERATE REST OF ARRAYS
310 FOR J=1 TO 100
311 Z4=W(J)*E(2,J)*E(2,J)
312 Z5=W(J)*E(2,J)*R(J)
313 Z7=W(J)*E(1,J)*E(2,J)
314 B(3,3)=B(3,3)+Z4
315 B(4,4)=B(4,4)+T(J)*T(J)*Z4
316 B(1,3)=B(1,3)+Z7
317 B(1,4)=B(1,4)+T(J)*Z7
318 B(2,4)=B(2,4)+T(J)*T(J)*Z7
319 B(3,4)=B(3,4)+T(J)*Z4
320 C(3)=C(3)+Z5
321 C(4)=C(4)+T(J)*Z5
322 NEXT J
323 B(2,3)=B(1,4)
324 Z6=A(3)/(A(4)*A(4))
325 B(4,4)=B(4,4)*Z6*Z6
326 B(1,4)=B(1,4)*Z6
327 B(2,3)=B(2,3)*Z3
328 B(2,4)=B(2,4)*Z3*Z6
329 B(3,4)=B(3,4)*Z6
330 C(4)=C(4)*Z6
331 ! TAKE ADVANTAGE OF SYMMETRY OF B MATRIX
332 FOR J=1 TO N @ FOR K=2 TO N
333 B(K,J)=B(J,K)
334 NEXT K @ NEXT J
335 RETURN
336 !
337 ! *******************************************************************
338 ! *                     EVALUATE CHI SQUARE                       *
339 ! *******************************************************************
340 ! RETURNS WITH CURRENT CHI SQUARE IN X2 AND RESIDUALS IN R()
341 ! EXP(-T(J)/A(2)) IN E(1,)
342 ! EXP(-T(J)/A(4)) IN E(2,)
```

```
343 ! FAST EXPONENTIAL EVALUATION ACCELERATES CALCULATIONS
344 X2=0
345 FOR K=1 TO N1 @ E(K,1)=1 @ Z=EXP(-T/A(2*K))
346 FOR J=2 TO 100 @ E(K,J)=E(K,J-1)*Z
347 NEXT J
348 NEXT K
349 FOR J=1 TO 100
350 Z=0
351 FOR K=1 TO N1
352 Z=Z+A(-1+2*K)*E(K,J)
353 NEXT K
354 R(J)=Z-D(J)
355 X2=X2+W(J)*R(J)*R(J)
356 NEXT J
357 RETURN
358 ! *****************************************************************
359 ! *                    GENERATE SYNTHETIC DATA                    *
360 ! *****************************************************************
361 ! USES FAST EXPONENTIAL EVALUATION
362 DISP "NUMBER OF COMPONENTS' IN SYNTHETIC DATA (1 OR 2)";
363 INPUT N2
364 T(1)=0 @ D(1)=A1 @ Z1=EXP(-T/A2) @ ! CONSTANTS FOR ITERATION
365 FOR J=2 TO 100
366 T(J)=(J-1)*T
367 D(J)=D(J-1)*Z1
368 NEXT J
369 IF N2=1 THEN 377
370 ! DO SAME FOR SECOND EXPONENTIAL
371 Z2=EXP(-T/A4) @ Z3=A3
372 D(1)=D(1)+A3
373 FOR J=2 TO 100
374 Z3=Z3*Z2
375 D(J)=D(J)+Z3
376 NEXT J
377 DISP "ADD NOISE TO DATA (Y OR N)";@ INPUT A$
378 IF A$="N" THEN RETURN
379 ! ADD NOISE USING BOX-MULLER-MARSAGLIA ALGORITHM (SEE TEXT)
380 FOR J=1 TO 100
381 Z1=2*RND-1 @ Z2=2*RND-1
382 S=Z1*Z1+Z2*Z2
383 IF S>1 THEN 381
384 Z3=SQR(-2*LOG(S)/S)
385 Z4=Z1*Z3
386 D(J)=D(J)+Z4*SQR(D(J))
387 NEXT J
388 RETURN
389 !
390 ! *****************************************************************
391 ! *                      PLOT RESIDUALS                           *
392 ! *****************************************************************
393 ! CALCULATE UNWEIGHTED OR WEIGHTED RESIDUALS
394 FOR J=1 TO 100 @ R(J)=SQR(W(J))*R(J) @ NEXT J
395 Z1=AMIN(R) @ Z2=AMAX(R) @ ! FIND LIMITS ON PLOT
396 Z3=ABS(Z1) @ Z4=ABS(Z2)
397 IF Z3>Z4 THEN MO=Z3 ELSE MO=Z4
398 ! SCALE PLOTTING AREA, ADD AXES AND LABEL
399 GCLEAR @ SCALE -40,110*T,-1.4*MO,1.1*MO
400 CLIP 0,100*T,-MO,MO
401 IF 100*T>10 THEN FXD 0,3 ELSE FXD 3,3 @ ! SET AXES DECIMALS
402 LAXES 10*T,MO/10,0,0,2,2,5
403 ! PLOT RESIDUALS
404 PENUP @ FOR J=1 TO 100 @ PLOT (J-1)*T,R(J) @ NEXT J
405 RETURN
```

```
101 ! ***********************************************************************
102 ! *                                                                    *
103 ! *                          SIMPLEX                                   *
104 ! *                                                                    *
105 ! ***********************************************************************
106 !
107 ! SIMPLEX FITTING OF DATA TO A K[EXP(-T/T1)-EXP(-T/T2)]
108 ! USER DEFINABLE FUNCTIONS ARE EASILY ADDED
109 ! FINAL BY J.N.DEMAS (1/19/83)
110 ! ADAPTED WITH PERMISSION FROM FORTRAN PROGRAM BY DANIELS (1978)
111 ! EARLY VERSIONS BY J.M.GREER AND B.HAUENSTEIN
112 ! WITH GRAPHICS
113 ! SOFTWARE DEFINED SPECIAL FUNCTION KEY SETS FLAG DURING RUNNING
114 ! TO PERMIT INTERRUPTING EXECUTION TO DISPLAY CURRENT BEST FIT
115 ! MAXIMUM #OF PARAMETERS IS 8
116 !
117 !        DEFINITION OF VARIABLES
118 ! I1=ITERATION COUNTER
119 ! E5=MINIMUM RELATIVE ERROR BETWEEN CHI SQU BEFORE DISPLAY
120 ! E9=FRACTIONAL CHANGE IN CHI SQUARES ON SUCCESSIVE ITERATIONS
121 ! B()=RESIDUALS BETWEEN OBSERVED AND CALCULATED FUNCTION
122 ! P(,)=ARRAY OF POINTS IN SIMPLEX
123 !     1ST INDEX=# OF P VECTOR,
124 !     2ND=VALUE OF THAT PARAMETER
125 ! T()=ARRAY OF INDEPENDENT VARIABLE
126 ! D()=ARRAY OF DEPENDENT VARIABLE
127 ! D1()=ARRAY OF CURRENTLY CALCULATED D()
128 ! W()=ARRAY OF WEIGHTING FACTORS
129 ! E()=ARRAY OF CHI SQUARES WITH E(Q)=CHI SQUARE FOR P(Q, )
130 ! P(9,) AND P(10,)=TEMPORARY VECTOR STORAGE LOCATIONS
131 ! N=#PARAMETERS IN MODEL EQUATION
132 ! N1=#POINTS IN SIMPLEX
133 ! L=INDEX OF P VECTOR WITH LOWEST CHI SQUARE
134 ! H=INDEX OF ONE WITH HIGHEST
135 ! H2=INDEX OF ONE WITH NEXT HIGHEST CHI SQUARE
136 ! V=CHI SQUARE
137 ! C()=ARRAY OF CENTROID
138 ! S9=EVERY S9th POINT USED IN GENERATING CHI SQUARE AT BEGINNING
139 ! P0,P1=FIRST AND LAST POINT FOR FITTING
140 !
141 I1=0 @ F=0 @ S9=10
142 ON KEY# 1 GOSUB 325 @ ! SET DISPLAY FLAG F=1 ON PRESSING KEY#1
143 ! F FLAG TESTED PERIODICALLY. GENERATES A DISPLAY WHEN SET TO 1
144 ! MACHINES WITHOUT THESE KEYS CAN BE STOPPED, F SET TO 1 FROM
145 ! KEYBOARD, EXECUTION CONTINUED
146 CLEAR
147 DISP "SIMPLEX OPTIMIZATION TO FIT A DIFFERENCE OF 2 EXPONENTIALS"
148 DISP "PRESS KEY #1 AT ANY TIME TO"
149 DISP "ACCESS THE DATA DISPLAY ROUTINE" @ DISP
150 DISP "FIRST POINT IN FITTING";@ INPUT P0
151 DISP "LAST POINT IN FITTING";@ INPUT P1
152 DISP "RELATIVE ERROR BEFORE" @ DISP "DATA DISPLAY";@ INPUT E5
153 DISP "ADD NOISE TO DATA (Y or N)";@ INPUT A$
154 OPTION BASE 1
155 DIM P(10,8),T(100),D(100),D1(100),W(100),C(10),E(10),B(100)
156 COM X9
157 ! ZERO OUT MATRICES
158 MAT D=ZER@ MAT D1=ZER@ MAT B=ZER
159 ! IF REDUCED # OF POINTS USED IN EVALUATING CHI SQUARE
160 ! TEMPORARILY RELAX CONVERGENCE CRITERIA (SEE TEXT)
```

```
.61 IF S9=10 THEN E5=100*E5
.62 ! *********************************************************************
.63 ! *    GENERATION OR ENTRY DATA. ON RETURN DATA IN T(),D(),W()     *
.64 ! *********************************************************************
.65 GOSUB 328
.66 !
.67 !
.68 ! ******** CHANGE N IN NEXT LINE IF #PARAMETERS CHANGED *********
.69 N=3
.70 N1=N+1 @ !   # POINTS IN SIMPLEX ONE GREATER THAN # OF PARAMETERS
.71 DISP
.72 ! ******************* INPUT INITIAL GUESSES *********************
.73 DISP "FIT TO DIFFERENCE OF 2 EXPONENTIALS"
.74 DISP "DISPLAYS PREEXPONENTIAL FACTORS THEN 2 LIFETIMES"
.75 DISP "INITIAL GUESSES FOR EACH VARIABLE. MUST NOT BE ZERO"
.76 FOR I=1 TO N
.77 DISP "VARIABLE # ";@ DISP I;@ INPUT P(1,I)
.78 IF P(1,I)=0 THEN BEEP @ DISP "NO ZERO GUESSES " @ GOTO 177
.79 NEXT I
.80 I=1 @ GOSUB 281 @ E(1)=V
.81 E4=E(1)
.82 PRINT "ITER.     K            T1          T2      CHI SQR"
.83 PRINT USING "DDD,4(8D.3D,3X)" ; I1,P(1,1),P(1,2),P(1,3),E(1)
.84 !
.85 ! **************** GENERATE INITIAL SIMPLEX ******************
.86 FOR I=2 TO N+1
.87 FOR J=1 TO N
.88 IF I-1=J THEN P(I,J)=1.1*P(1,J) ELSE P(I,J)=P(1,J)
.89 NEXT J
.90 NEXT I
.91 !
.92 ! *********************************************************************
.93 ! *                    MAIN PROGRAM                               *
.94 ! *********************************************************************
.95 !
.96 ! ******* FIND POINTS WITH LOWEST ERROR AND HIGHEST ERROR *******
.97 L=1 @ H=1
.98 FOR I=1 TO N1 @ GOSUB 281 @ E(I)=V @ IF E(I)<E(L) THEN L=I
.99 IF E(I)>E(H) THEN H=I
200 NEXT I
201 ! ************** FIND POINT WITH NEXT LOWEST ERROR **************
202 H2=L @ FOR I=1 TO N1 @ IF E(I)>=E(H2) AND I<>H THEN H2=I
203 NEXT I
204 !
205 ! ****************** CALCULATE CENTROID *********************
206 FOR J=1 TO N
207 C(J)=-P(H,J)
208 FOR I=1 TO N1 @ C(J)=C(J)+P(I,J) @ NEXT I
209 C(J)=C(J)/N
210 NEXT J
211 !
212 ! ****************** REFLECT SIMPLEX   ********************
213 ! REFLECTED POINT PASSED TO CHI SQUARE ROUTINE IN P(10,)
214 FOR J=1 TO N @ P(10,J)=1.9985*C(J)-.9985*P(H,J) @ NEXT J
215 I=10 @ GOSUB 281 @ E1=V
216 !
217 ! ***************** IF VERY SUCCESSFUL EXPAND ****************
218 IF E1<E(L) THEN L=H @ H=H2 @ GOTO 232
219 !
220 ! ****************** IF UNSUCCESSFUL CONTRACT ***************
```

```
221 IF E1>=E(H) THEN 254
222 !
223 ! ************** IF MODERATELY SUCCESSFUL REFLECT AGAIN *********
224 FOR J=1 TO N @ P(H,J)=P(10,J) @ NEXT J
225 E(H)=E1
226 IF E1>E(H2) THEN 214
227 H=H2 @ GOTO 202
228 !
229 ! ********************** EXPAND SIMPLEX ************************
230 ! EXPANDED POINT PASSED TO CHI SQUARE ROUTINE IN P(9,)
231 !
232 FOR I=1 TO N @ P(9,I)=1.95*P(10,I)-.95*C(I) @ NEXT I
233 I=9 @ GOSUB 281 @ E2=V
234 ! MAKE BEST OF REFLECTED OR EXPANDED POINT NEW LOWEST
235 IF E2<E1 THEN 237
236 FOR J=1 TO N @ P(L,J)=P(10,J) @ NEXT J @ E(L)=E1 @ GOTO 241
237 FOR J=1 TO N @ P(L,J)=P(9,J) @ NEXT J @ E(L)=E2
238 !
239 ! *************** DISPLAY DATA WITH LOWEST ERROR **************
240 !    ROUTINE IS FOR 3 PARAMETERS, MODIFY FOR OTHER VALUES
241 I1=I1+1 @ REM INCREMENT ITERATION COUNTER
242 PRINT USING "DDD,4(8D.3D,3X)" ; I1,P(L,1),P(L,2),P(L,3),E(L)
243 ! BRANCH IF ERROR LIMITS REACHED
244 E9=ABS((E4-E(L))/E4) @ ! E9=DELTA CHANGE CHI SQU
245 ! IF CONVERGED & EVERY 10th POINT USED, TIGHTEN CONVERGENCE
246 ! AND USE ALL POINTS
247 IF E9>=E5 THEN 250
248 IF S9=10 THEN S9=1 @ E5=.01*E5 @ DISP "ALL POINTS" @ GOTO 197
249 IF E9<E5 THEN DISP "RELATIVE ERROR LIMIT REACHED" @ GOTO 258
250 E4=E(L)
251 IF F=1 THEN DISP "GRAPHICS DISPLAY" @ GOTO 258 ELSE 202
252 !
253 ! ********************* CONTRACT SIMPLEX **********************
254 FOR J=1 TO N @ P(10,J)=.5015*C(J)+.4985*P(H,J) @ NEXT J
255 J=10 @ GOSUB 281 @ E1=V @ IF E1>E(L) THEN 257
256 L=H @ H=H2 @ GOTO 236
257 IF E1<E(H) THEN 224
258 BEEP @ DISP "NEW SIMPLEX(1) OR A DATA DISPLAY(2)";@ INPUT O@ F=0
259 IF O=2 THEN 290
260 ! **************** SCALE IF YOU GOT TO HERE *******************
261 DISP "SCALING FACTOR";@ INPUT K@ DISP "NEW MIN. ERROR";@ INPUT E3
262 DISP "NEW RELATIVE ERROR";@ INPUT E5@ DISP
263 ! SCALE PLOTTING AREA
264 FOR I=1 TO N1 @ FOR J=1 TO N
265 P(I,J)=P(I,J)+K*(P(L,J)-P(I,J))
266 NEXT J @ NEXT I
267 GOTO 197 @ ! RESUME SIMPLEX REDUCTION WITH SCALED POINTS
268 !
269 ! ************************************************************
270 ! *                 ERROR EVALUATION ROUTINE                *
271 ! ************************************************************
272 ! RETURNS WITH CHI SQUARE IN V
273 !                 CALCULATED FIT IN D1()
274 !                 DIFFERENCES BETWEEN D() AND D1() IN B()
275 ! BECAUSE OF ABSENCE OF TRUE SUBROUTINES IN BASIC, PARAMETERS
276 ! ARE PASSED BY WRITING FUNCTION IN THE WAY SHOWN BELOW
277 ! P(I,1), P(I,2) AND P(I,3) IS USED FOR PARAMETERS 1, 2 AND 3
278 ! EVERY TIME THE ROUTINE IS CALLED,
279 ! I IS SET EQUAL TO THE INDEX OF THE VECTOR TO BE PASSED
280 !
```

```
281 V=0
282 FOR Z=PO TO P1 STEP S9
283 D1(Z)=P(I,1)*(EXP(-T(Z)/P(I,2))-EXP(-T(Z)/P(I,3)))
284 B(Z)=D(Z)-D1(Z)
285 V=V+W(Z)*B(Z)*B(Z)
286 NEXT Z
287 RETURN
288 !
289 ! ****************************************************************
290 ! *                    DATA DISPLAY ROUTINE                     *
291 ! ****************************************************************
292 GCLEAR
293 ! SET LIMITS ON AXIS
294 U1=AMAX(D) @ U2=AMAX(D1)
295 IF U1<U2 THEN M2=U2 ELSE M2=U1
296 M1=T(P1) @ ! M1= MAX TIME
297 SCALE -.3*M1,1.1*M1,-.45*M2,1.1*M2 @ ! SET GRAPHICS FIELD
298 CLIP 0,M1,0,M2
299 FXD 3,0 @ ! SET DECIMALS ON AXIS NUMERATION
300 LAXES M1/10,M2/10,0,0,2,2,5 @ ! DRAW NUMBER AXES
301 ! PLOT CALCULATED AND OBSERVED DECAYS
302 PENUP @ FOR Z=1 TO P1 @ PLOT T(Z),D1(Z) @ NEXT Z
303 PENUP @ FOR Z=1 TO P1 @ PLOT T(Z),D(Z) @ NEXT Z
304 ! PUT INSTRUCTIONS ON SCREEN
305 MOVE .1*M1,.8*M2
306 LABEL "AFTER BEEP, PRESS CONT" @ LABEL "TO PLOT ERROR FUNCTION"
307 ! FIND SCALING FOR B PLOT
308 B1=AMAX(B) @ B2=ABS(AMIN(B))
309 IF B2>B1 THEN B1=B2
310 BEEP @ PAUSE
311 ! PLOT RESIDUALS
312 GCLEAR @ SCALE -.4*M1,1.1*M1,-1.8*B1,1.1*B1 @ ! SET GRAPH FIELD
313 CLIP 0,M1,-1.001*B1,B1
314 FXD 3,2 @ ! FIX DECIMALS ON AXIS AND PLOT THEM
315 LAXES M1/10,.25*B1,0,0,2,2,5 @ ! DRAW-LABEL AXES
316 FOR Z=1 TO P1 @ PLOT T(Z),B(Z) @ NEXT Z
317 ! PUT UP INSTRUCTIONS
318 MOVE .2*M1,.9*B1
319 LABEL "ERROR FUNCTION"
320 LABEL "PRESS CONTINUE" @ LABEL "FOR MORE DATA" @ LABEL "REDUCTION"
321 ! CLEAR NEED TO PLOT FLAG
322 BEEP @ PAUSE @ ! ADMIRE PLOT
323 F=0 @ GOTO 202 @ ! MORE DATA REDUCTION
324 !
325 ! ***** SOFTWARE SET FLAG TO SIGNAL THAT A DISPLAY IS WANTED ****
326 ! SET WHEN KEY#1 IS PRESSED
327 F=1 @ RETURN
328 !
329 ! ****************************************************************
330 ! *                   DATA GENERATION ROUTINE                   *
331 ! ****************************************************************
332 ! DIFFERENCE OF 2 EXPONENTIALS WITH LIFETIMES OF 35 AND 7
333 ! SCALED TO 10000 PEAK
334 FOR Z=1 TO P1
335 Z1=Z-1 @ T(Z)=Z1
336 D(Z)=EXP(-T(Z)/35)-EXP(-T(Z)/7)
337 NEXT Z
338 Z1=1/AMAX(D) @ ! FIND SCALING FACTOR FOR PEAK ADJUSTMENT
339 FOR Z=1 TO P1 @ D(Z)=10000*Z1*D(Z) @ NEXT Z
340 ! ***** GENERATE WEIGHTING FACTOR- IN THIS PROGRAM ALL 1'S ******
```

```
341 FOR Z=1 TO 100 @ W(Z)=1 @ NEXT Z
342 IF A$#"Y" THEN RETURN
343 !
344 ! ****** NOISE ADDITION BY BOX-MULLER-MARSAGLIA ALGORITHM *******
345 ! STANDARD DEVIATION =SQUARE ROOT OF CHANNEL COUNTS
346 ! Z4= NORMAL DEVIATE WITH MEAN OF 0 & STANDARD DEVIATION OF 1
347 FOR J=1 TO P1
348 Z1=2*RND-1 @ Z2=2*RND-1
349 S=Z1*Z1+Z2*Z2
350 IF S>1 THEN 348
351 Z3=SQR(-2*LOG(S)/S)
352 Z4=Z1*Z3
353 D(J)=D(J)+Z4*SQR(D(J))
354 NEXT J
355 DISP "NOISE ADDED" @ RETURN
356 END
```

```
101 ! *****************************************************************
102 ! *                                                               *
103 ! *                         LUMEN                                 *
104 ! *                                                               *
105 ! *****************************************************************
106 ! WRITTEN BY A.T. WYKER REVISED BY J.N. DEMAS (MODIFIED 1/1/83)
107 ! SEE DEMAS(1976); WYKER-DEMAS(1976); GREER,REED,DEMAS(1981)
108 ! RUNS WITH EXTERNAL PRINTER BECAUSE OF VOLUMINOUS OUTPUT
109 !        VARIABLE USAGE
110 !    E()=OBSERVED EXCITATION
111 !    D()=OBSERVED DECAY
112 !    X()=CALCULATED DECAY
113 !    W() AND Z()=NORMAL PHASE PLANE VARIABLES
114 !    R() AND Q() =NORMAL SUMS USED TO CALCULATE W() AND Z()
115 !    T=INTERVALS BETWEEN POINTS
116 !    C=NUMBER OF DATA POINTS
117 !    G2=SUM OF SQUARES OF RESIDUALS
118 !    T1=CALCULATED LIFETIME
119 !    K1=CALCULATED PROPORTIONALITY CONSTANT
120 !    Z9=FLAG: FIRST(1) OR SUBSEQUENT(2) PASSES THROUGH PROGRAM
121 !    F=FLAG: KEYBOARD(1) OR DATA STATEMENT(2) INPUT
122 ! SAMPLE DATA APPENDED. IF FIT FROM POINT 5 ON YIELDS:
123 ! LIFETIME OF 173.02
124 ! K OF .006332295
125 ! REDUCED CHI SQUARE OF 1.06174
126 ! CALCULATED VALUE AT T=184.4 OF 4.25441135
127 ! COM USED TO REDUCE PROGRAM STORAGE SPACE-HP QUIRK
128 COM X9
129 OPTION BASE 1
130 CLEAR @ REM CLEARS DISPLAY
131 DISP "LUMEN-PROGRAM FOR DECONVOLUTION OF LUMINESCENCE LIFETIMES"
132 DISP "USES PHASE PLANE METHOD"
133 DISP @ DISP "ACCEPTS <51 POINTS"
134 DISP "PERMITS MULTIPLE DECAYS TO BE USED WITH A SINGLE FLASH"
135 DISP "USES GRAPHICS- AFTER EACH PLOT THERE IS A BEEP"
```

```
36 DISP "TO COPY PLOT, PRESS SHIFT-COPY TO COPY."
37 DISP "PRESS CONT TO CONTINUE CALCULATIONS- INCLUDING FROM HERE"
38 BEEP @ PAUSE
39 CLEAR
40 DISP @ DISP "ALL YES/NO QUESTIONS ARE ANSWERED BY 'YES' OR 'NO'"
41 DISP
42 PRINTER IS 10,132 @ ! SET UP PRINTER
43 Z9=0 @ ! SET FIRST PASS FLAG
44 DISP "DATA MAY BE ENTERED BY KEYBOARD(1) OR DATA STATEMENTS(2)"
45 DISP "WHICH ONE? (i.e. 1 OR 2)";@ INPUT F
46 IF F#1 AND F#2 THEN 144
47 DIM E(50),D(50),Z(50),W(50),X(50),A$[3],B$[3]
48 IF Z9=0 THEN 151
49 DISP @ DISP "DO YOU WISH TO INPUT A NEW FLASH? (YES OR NO) ";
50 INPUT A$@ IF A$="NO" THEN 158
51 IF F=1 THEN DISP @ DISP "INPUT TIME INTERVAL BETWEEN POINTS ";
52 IF F=1 THEN INPUT T ELSE READ T
53 IF F=1 THEN DISP "INPUT # READINGS TAKEN-INCLUDE ZERO READING";
54 IF F=1 THEN INPUT C ELSE READ C
55 REDIM E(C),D(C),Z(C),W(C),X(C)
56 IF F=1 THEN DISP @ DISP "INPUT E(T)'S. BEGIN WITH ZERO READING ";
57 IF F=1 THEN MAT INPUT E ELSE MAT READ E
58 IF F=1 THEN DISP @ DISP "INPUT D(T)'S. BEGIN WITH ZERO READING";
59 IF F=1 THEN INPUT D ELSE MAT READ D
60 !
61 ! ****************************************************************
62 ! *              DETERMINE W(T) AND Z(T)                        *
63 ! ****************************************************************
64 ! ZERO OUT ALL SUM TERMS BEFORE BEGINNING
65 A=0 @ B=0 @ G=0 @ E=0 @ G2=0 @ J6=0 @ R=0 @ Q=0 @ W(1)=0 @ Z(1)=0
66 FOR J=2 TO C
67 R=D(J)+D(J-1)+R
68 Q=E(J)+E(J-1)+Q
69 W(J)=2*D(J)/(T*Q)
70 Z(J)=R/Q
71 NEXT J
72 GOSUB 236 @ REM PLOT PHASE PLANE PLOT
73 !
74 ! ****************************************************************
75 ! *          LEAST SQUARES FIT OF W(T) VS Z(T)                  *
76 ! *          FIT MAY BEGIN AT ANY POINT CALCULATION             *
77 ! ****************************************************************
78 DISP @ DISP "FIRST POINT OF PP PLOT TO BEGIN FITTING";
79 INPUT P@ IF P>=C THEN 178
80 FOR J=P TO C
81 A=A+W(J)*Z(J)
82 B=B+Z(J)
83 E=E+W(J)
84 G=G+Z(J)*Z(J)
85 J6=J6+1
86 NEXT J
87 D1=G*J6-B*B
88 S=(A*J6-B*E)/D1 @ ! CALC SLOPE OF LEAST SQUARES LINE
89 K1=(E*G-B*A)/D1 @ ! CALC INTERCEPT PP PLOT
90 !
91 ! ************* DRAW IN LEAST SQUARES LINE *********************
92 MOVE 0,K1
93 DRAW U1,K1+S*U1
94 BEEP @ PAUSE @ ! WAIT FOR OPERATOR TO CONTINUE
95 T1=-1/S
```

```
196 PRINT " LIFETIME= ";T1
197 PRINT "K VALUE =   ";K1
198 !
199 ! ****************************************************************
200 ! *     CALCULATE D(CALC) USING DEMAS-CROSBY(1970) ITERATIVE    *
201 ! *       CONVOLUTION FORMULA TO MINIMIZE # OF POINTS NEEDED    *
202 ! ****************************************************************
203 E9=EXP(-T/T1) @ X(1)=0 @ REM INITIALIZE VARIABLES
204 FOR K=2 TO C
205 L1=(E(K)-E(K-1))/T
206 N1=E(K)-L1*(K-1)*T
207 U=L1*((K-1)*T-T1)+N1
208 V=E9*(L1*(T1-(K-2)*T)-N1)
209 X(K)=(U+V)*K1*T1+X(K-1)*E9
210 D=X(K)-D(K)
211 G2=G2+D*D
212 NEXT K
213 ! *************** PRINT OUT RESULTS*******************************
214 PRINT "REDUCED CHI SQUARE (UNIT WEIGHTS ASSUMED)= ";G2/(C-2)
215 PRINT "SQUARE ROOT REDUCED CHI SQUARE = ";SQR(G2/(C-2))
216 GOSUB 258 @ ! PLOT DECAYS
217 DISP @ DISP "DO YOU WANT THE VALUES PRINTED";
218 INPUT A$@ IF A$="NO" THEN 225
219 IF A$="NO" THEN 225
220 PRINT @ PRINT
221 PRINT "TIME","W(T)","Z(T)","FLASH","D(OBS)","DCALC"
222 FOR L3=1 TO C
223 PRINT (L3-1)*T,W(L3),Z(L3),E(L3),D(L3),X(L3)
224 NEXT L3
225 DISP @ DISP "DO YOU WANT TO CONTINUE";
226 INPUT B$@ IF B$="NO" THEN 229
227 Z9=1 @ REM SET COMPLETED A PASS FLAG
228 GOTO 148
229 CLEAR @ DISP "FINI. HAVE A NICE DAY" @ END
230 !
231 ! ****************************************************************
232 ! *                 PLOT PHASE PLANE PLOT                       *
233 ! ****************************************************************
234 !    FIND MIN AND MAX FOR SCALING OF PLOTTER AREA
235 U1=AMAX(Z) @ U2=AMAX(W)
236 GCLEAR
237 !    SET UP DECIMALS ON LABEL NUMBERS THEN DRAW LABEL AXIS
238 F8=4-INT(LGT(U1)) @ IF F8>4 THEN F8=4
239 F9=4-INT(LGT(U2)) @ IF F9>5 THEN F9=5
240 FXD F8,F9
241 SCALE -.3*U1,1.1*U1,-.4*U2,1.2*U2
242 CLIP 0,U1,0,U2
243 LAXES U1/10,U2/10,0,0,2,2,5
244 MOVE U1/2,1.2*U2
245 LORG 6 @ LABEL USING "K" ; "W(T) VS Z(T)"
246 FOR J=2 TO C
247 MOVE Z(J),W(J)
248 LORG 5 @ LABEL USING "K" ; "*"
249 IF 5*INT(J/5)=J THEN LORG 1 @ MOVE Z(J),W(J) @ LABEL USING "K" ; J
250 NEXT J
251 ALPHA @ RETURN
252 !
253 ! ****************************************************************
254 ! *            PLOT E(T),D(T), AND BEST FIT "*"                 *
255 ! ****************************************************************
```

```
256 !
257 ! ************* FIND MAXIMA OF FUNCTIONS ************************
258 M4=AMAX(E) @ M5=AMAX(D) @ M6=AMAX(X)
259 C1=C-1
260 GCLEAR @ SCALE -.2*T*C1,1.1*T*C1,-40,120 @ MOVE T*C1/2,120
261 LORG 6 @ LABEL USING "K" ; "E(T),D(T), BEST FIT(*)"
262 CLIP 0,T*C1,0,100
263 ! ADJUST DECIMALS ON LABEL AXIS THEN DRAW ,NUMBER AXIS
264 F8=4⌐INT(LGT(T*C1)) @ IF F8>5 THEN F8=5
265 FXD F8,0
266 LAXES T*C1/10,10,0,0,2,2,5 @ ! DRAW AXES
267 ! ********************** PLOT E(T)***************************
268 FOR J=1 TO C
269 PLOT T*(J-1),100*E(J)/M4
270 NEXT J
271 IF M6>M6 THEN N=M5 ELSE N=M5 @ REM SELECT SCALE FOR (D)
272 PENUP
273 !
274 ! *********************** PLOT D(T) **********************
275 FOR J=1 TO C
276 PLOT T*(J-1),100*D(J)/N
277 NEXT J
278 LORG 5 @ REM SET LABEL ORIGIN TO CENTER *'S ON DATA POINTS
279 !
280 ! *************** POINT PLOT BEST FIT ***********************
281 FOR J=1 TO C
282 MOVE T*(J-1),100*X(J)/N
283 LABEL USING "K" ; "*"
284 NEXT J
285 BEEP @ PAUSE
286 ALPHA @ RETURN
287 !
288 ! ***************** SAMPLE DATA ***************************
289 ! TIME INTERVAL THEN NUMBER OF DATA POINTS
290 DATA 92.2,26
291 ! FLASH DATA
292 DATA 0,3,13,33,69,101,129,150,156
293 DATA 156,148,136,124,111,98,88,78,71,64,58,52,48,43,39,36,34
294 ! DECAY DATA
295 DATA 0,1,4,14,31,56,85,114,139,151,156
296 DATA 156,152,142,130,119,108,95,86,78,71,65,57,52,47,44
```

References

Almgren, M., Grieser, F., and Thomas, J. K. (1979). *J. Am. Chem. Soc.* **101**, 2021.
Andre, J. C., Vincent, L. M., O'Connor, D. V., and Ware, W. R. (1979). *J. Phys. Chem.* **83**, 2285.
Aoki, T., and Sakurai, K. (1980). *Rev. Sci. Instrum.* **51**, 951.
Atik, S. S., and Thomas, J. K. (1981). *J. Am. Chem. Soc.* **103**, 3550.
Bacon, J. R., and Demas, J. N. (1983). *Anal. Chem.* **55**, 653.
Badea, M. G., and Georghiou, S. (1976). *Rev. Sci. Instrum.* **47**, 314.
Balzani, V., Moggi, L., Manfrin, M. F., and Bolletta, F. (1975). *Coord. Chem. Rev.* **15**, 321.
Bauer, R. K., Borenstein R., de Mayo, P., Okada, K., Rafalska, M., Ware, W. R., and Wu, C. (1982a). *J. Am. Chem. Soc.* **104**, 4635.
Bauer, R. K., de Mayo, P., Ware, W. R., and Wu, K. C. (1982b). *J. Phys. Chem.* **86**, 3781.
Baxendale, J. H., and Rodgers, M. A. J. (1980). *Chem. Phys. Lett.* **72**, 424.
Bay, Z. (1950). *Phys. Rev.* **77**, 419.
Beavan, S. W., Hargreaves, J. S., and Phillips, D. (1979). *Adv. Photochem.* **11**, 207.
Beers, Y. (1962). "Introduction To The Theory Of Error," 2nd ed. Addison-Wesley, Reading, Massachusetts.
Bennett, R. G. (1964). *J. Chem. Phys.* **41**, 3037.
Bennett, R. G., Schwenker, R. P., and Kellogg, R. E. (1964). *J. Chem. Phys.* **41**, 3040.
Benson, S. W. (1960). "The Foundations of Chemical Kinetics." McGraw-Hill, New York.
Berlman, I. B. (1971). "Handbook of Fluorescence Spectra of Aromatic Molecules." Academic Press, New York.
Bevington, P. R. (1969). "Data Reduction and Error Analysis For the Physical Sciences." McGraw-Hill, New York.
Birks, J. B. (1970). "Photophysics of Aromatic Molecules." Wiley (Interscience), New York.
Birks, J. B., and Christophorov, L. G. (1963). *Spectrochim. Acta* **19**, 401.
Birks, J. B., and Munro, I. H. (1967). *Prog. React. Kinet.* **4**, 239.
Birks, J. B., Dyson, D. J., and Munro, I. H. (1963). *Proc. R. Soc. London, Ser. A* **275**, 575.
Borders, R. A., Birks, J. W., and Borders, J. A. (1980). *Anal. Chem.* **52**, 1273.
Box, G. E. P., and Muller, M. E. (1958). *Ann. Math. Stat.* **29**, 610.
Bridgett, K. A., King, T. A., and Smith-Saville, R. J. (1970). *J. Phys. E* **3**, 767.
Brittain, H. G. (1979). *Inorg. Chem.* **18**, 1740.
Brody, S. S. (1957). *Rev. Sci. Instrum.* **28**, 1021.
Caldwell, R. A., and Cao, C. V. (1981). *J. Am. Chem. Soc.* **103**, 3594.

Cantor, C. R., and Tao, T. (1971). *In* "Procedures In Nucleic Acid Research" (G. L. Cantoni, ed.), Vol. 2, p. 31. Harper, New York.

Capellos, C., and Bielski, B. H. J. (1972). "Kinetic Systems." Wiley (Interscience), New York.

Carter, T. P., and Gillispie, G. D. (1980). *Chem. Phys. Lett.* **73**, 75.

Chakrabarti, S. K., and Ware, W. R. (1971). *J. Chem. Phys.* **55**, 5494.

Choi, J. D., Fugate, R. D., and Song, P.-S. (1980). *J. Am. Chem. Soc.* **102**, 5293.

Clayton, R. K. (1971). "Light and Living Matter," Vol. 2. McGraw-Hill, New York.

Coates, P. B. (1968). *J. Phys. E* **1**, 878.

Coates, P. B. (1972). *J. Phys. E* **5**, 148.

Cooper, D. H. (1966). *Rev. Sci. Instrum.* **37**, 1407.

Cramer, L. E., and Spears, K. G. (1978). *J. Am. Chem. Soc.* **100**, 221.

Creutz, C., Chou, M., Netzel, T. L., Okumura, M., and Sutin, N. (1980). *J. Am. Chem. Soc.* **102**, 1309.

Crosby, G. A. (1975). *Acc. Chem. Res.* **8**, 231.

Crosby, G. A., Demas, J. N., and Callis, J. B. (1972). *J. Res. Natl. Bur. Stand., Sect. A* **76A**, 561.

Csorba, I. P. (1980). *Appl. Opt.* **19**, 3863.

Cundall, R. B., and Dale, R. E., eds. (1983). "Time-Resolved Fluorescence Spectroscopy in Biochemistry and Biology." Plenum, New York.

Daniels, R. W. (1978). "An Introduction to Numerical Methods and Optimization Techniques." North-Holland Publ., Amsterdam.

Davis, C. C., and King, T. A. (1970). *J. Phys. A* **3**, 101.

DeArmond, M. K. (1974). *Acc. Chem. Res.* **7**, 309.

DeGraff, B. A., and Demas, J. N. (1980). *J. Am. Chem. Soc.* **102**, 6169.

Demas, J. N. (1976). *J. Chem. Educ.* **53**, 657.

Demas, J. N. (1982). *In* "Photoluminescence Spectrometry" (K. D. Mielenz, ed.), p. 195. Academic Press, New York.

Demas, J. N., and Adamson, A. W. (1971). *J. Phys. Chem.* **75**, 2463.

Demas, J. N., and Adamson, A. W. (1973). *J. Am. Chem. Soc.* **95**, 5159.

Demas, J. N., and Addington, J. W. (1974). *J. Am. Chem. Soc.* **97**, 3838.

Demas, J. N., and Addington, J. W. (1976). *J. Am. Chem. Soc.* **98**, 5800.

Demas, J. N., and Crosby, G. A. (1970). *Anal. Chem.* **42**, 1010.

Demas, J. N., and Crosby, G. A. (1971). *J. Phys. Chem.* **75**, 991.

Demas, J. N., Addington, J. W., Peterson, S. H., and Harris, E. W. (1977a). *J. Phys. Chem.* **81**, 1039.

Demas, J. N., Harris, E. W., and McBride, R. P. (1977b). J. Am. Chem. Soc. **99**, 3547.

DeToma, R. P. (1983). *In* "Time-Resolved Fluorescence Spectroscopy in Biochemistry and Biology" (R. B. Cundall and R. E. Dale, eds.), p. 393. Plenum, New York.

Donnelly, V. M., and Kaufman, F. (1977). *J. Chem. Phys.* **66**, 4100.

Easter, J. H., DeToma, R. P., and Brand, L. (1976). *Biophys. J.* **16**, 571.

Eisenfeld, J. (1979). *IEEE Trans. Autom. Control* **24**, 498.

Eisenfeld, J. (1983). *In* "Time-Resolved Fluorescence Spectroscopy in Biochemistry and Biology" (R. B. Cundall and R. E. Dale, eds.), pp. 223, 233. Plenum, New York.

Eisenfeld, J., and Ford, C. C. (1979). *Biophys. J.* **26**, 73.

Elfring, W. H., Jr., and Crosby, G. A. (1981). *J. Am. Chem. Soc.* **103**, 2683.

Escabi-Perez, J. R., and Fendler, J. H. (1978). *J. Am. Chem. Soc.* **100**, 2234.

Förster, Th. (1948). *Ann. Phys. (Leipzig)* **2**, 55.

Förster, Th. (1949). *Z. Naturforsch. A,* **4A**, 321.

Förster, Th. (1960). *In* "Comparative Effects of Radiation" (M. Burton, J. S. Kirby-Smith, and J. L. Magee, eds.), p. 300. Wiley, New York.

Friedlander, G., Kennedy, J. W., and Miller, J. M. (1964). "Nuclear and Radiochemistry," 2nd ed. Wiley, New York.

Frost, A. A., and Pearson, R. G. (1953). "Kinetics and Mechanism." Wiley, New York.

Gafni, A., Modlin, R. L., and Brand, L. (1975). *Biophys. J.* **15**, 263.

Gibbons, J. D. (1976). "Nonparametric Methods for Quantitative Analysis." Holt, New York.

Gibson, E. P., and Rest, A. J. (1980). *Chem. Phys. Lett.* **73**, 294.

Goeringer, D. E., and Pardue, H. L. (1979). *Anal. Chem.* **51**, 1054.

Grätzel, M. (1981). *Acc. Chem. Res.* **14**, 376.

Grätzel, M., and Thomas, J. K. (1976). "Modern Fluorescence Spectroscopy," Vol. 2. Plenum, New York.

Greer, J. M., Reed, F. W., and Demas, J. N. (1981). *Anal. Chem.* **53**, 710.

Grinvald, A. (1976). *Anal. Biochem.* **75**, 260.

Grinvald, A., and Steinberg, I. Z. (1974). *Anal. Biochem.* **59**, 583.

Haar, H. P., and Hauser, M. (1978). *Rev. Sci. Instrum.* **49**, 632.

Haar, H. P., Klein, U. K. A., Hafner, F. W., and Hauser, M. (1977). *Chem. Phys. Lett.* **49**, 563.

Hallmark, J., and Eisenfeld, J. (1979). "Applied Nonlinear Analysis." Academic Press, New York.

Hamamatsu Corp. (1978). Technical Data Sheet, "Temporaldisperser," C979. Middlesex, New Jersey.

Hammond, P. R. (1979). *J. Chem. Phys.* **70**, 3884.

Harrigan, R. W., and Crosby, G. A. (1973). *J. Chem. Phys.* **59**, 3468.

Harris, C. M., and Selinger, B. K. (1979). *Aust. J. Chem.* **32**, 2111.

Harris, C. M., and Selinger, B. K. (1980). *J. Phys. Chem.* **84**, 891.

Harris, J. M., Chrisman, R. W., and Lytle, F. E. (1975). *Appl. Phys. Lett.* **26**, 16.

Harris, J. M., Lytle, F. E., and McCain, T. C. (1976). *Anal. Chem.* **48**, 2095.

Hartig, P. R., and Sauer, K. (1976). *Rev. Sci. Instrum.* **47**, 1122.

Hauenstein, B. L., Jr., Dressick, W. J., Buell, S. L., Demas, J. N., and DeGraff, B. A. (1983a). *J. Am. Chem. Soc.* **105** (in press).

Hauenstein, B. L., Mandal, K., Demas, J. N., and DeGraff, B. A. (1983b). *J. Phys. Chem.* (submitted).

Hauser, A. (1965). "Introduction to the Principles of Mechanics." Addison-Wesley, Reading, Massachusetts.

Hautala, R. R., King, R. B., and Kutal, C., eds. (1979). "Solar Energy Chemical Conversion and Storage." Humana Press, Clifton, New Jersey.

Hazan, G., Grinvald, A., Maytal, M., and Steinberg, I. Z. (1974). *Rev. Sci. Instrum.* **45**, 1602.

Helman, W. P. (1971). *Int. J. Radiat. Phys. Chem.* **3**, 283.

Hinde, A. L., Selinger, B. K., and Nott, P. R. (1977). *Aust. J. Chem.* **30**, 2383.

Hipps, K. W., and Crosby, G. A. (1975). *J. Am. Chem. Soc.* **97**, 7042.

Horrocks, W. D., Jr., and Sudnick, D. R. (1979). *J. Am. Chem. Soc.* **101**, 334.

Hunter, T. F., and Stock, M. G. (1974). *J.C.S. Faraday II* **70**, 1028.

Imhof, R. E., and Read, F. H. (1977). *Rep. Prog. Phys.* **40**, 1.

Irvin, J. A., Quickenden, T. I., and Sangster, D. F. (1981). *Rev. Sci. Instrum.* **52**, 191.

Isenberg, I. (1973). *J. Chem. Phys.* **59**, 5696.

Isenberg, I., and Dyson, R. D. (1969). *Biophys. J.* **9**, 1337.

Isenberg, I., Dyson, R. D., and Hanson, R. (1973). *Biophys. J.* **13**, 1090.

Jezequel, J. Y., Bouchy, M., and Andre, J. C. (1982). *Anal. Chem.* **54**, 2199.

Johnson, D. W., Christian, G. D., and Callis, J. B. (1977). *Anal. Chem.* **49**, 747A.

Johnson, J. K. (1980). "Numerical Methods in Chemistry." Dekker, New York.

Kalyanasundaram, K. (1978). *Chem. Soc. Rev,* **7**, 453.

Kalyanasundaram, K. (1982). *Coord. Chem. Rev.* **46**, 159.

Kaplan, W. (1952). "Advanced Calculus." Addison-Wesley, Reading, Massachusetts.

Kaplan, W. (1962). "Operational Methods for Linear Systems." Addison-Wesley, Reading, Massachusetts.

Kelder, S., and Rabani, J. (1981). *J. Phys. Chem.* **85**, 1637.

Kinoshita, S., and Kushida, T. (1982). *Rev. Sci. Instrum.* **53**, 469.

Kinoshita, S., Hironabu, O., and Kushida, T. (1981). *Rev. Sci. Instrum.* **52**, 572.

Kleinerman, M., and Choi, S.-I. (1968). *J. Chem. Phys.* **49**, 3901.

Knight, A. E. W., and Selinger, B. K. (1971). *Spectrochim. Acta, Part A* **27A**, 1223.

Knight, A. E. W., and Selinger, B. K. (1973). *Aust. J. Chem.* **26**, 1.

Knuth, D. E. (1969). "Seminumerical Algorithms: The Art of Computer Programming," Vol. 2. Addison-Wesley, Reading, Massachusetts.

Kobayashi, T., and Ohashi, Y. (1982). *Chem. Phys. Lett.* **86**, 289.

Koester, V. J. (1979). *Anal. Chem.* **41**, 458.

Kouyama, T. (1978). *Jpn. J. Appl. Phys.* **17**, 1409.

Kropp, J. L., and Windsor, M. W. (1963). *J. Chem. Phys.* **39**, 2769.

Kropp, J. L., and Windsor, M. W. (1965). *J. Chem. Phys.* **42**, 1599.

Kropp, J. L., and Windsor, M. W. (1966). *J. Chem. Phys.* **45**, 761.

Krug, W. P., and Demas, J. N. (1979). *J. Am. Chem. Soc.* **101**, 4394.

Laidler, K. J. (1965). "Chemical Kinetics," 2nd ed. McGraw-Hill, New York.

Lakowicz, J. R. (1981). *In* "Spectroscopy in Biochemistry" (J. Ellis Bell, ed.), Vol. 1, p. 195. CRC Press, Cleveland, Ohio.

Lakowicz, J. R., and Cherek, H. (1981a). *J. Biol. Chem.* **256**, 6348.

Lakowicz, J. R., and Cherek, H. (1981b). *J. Biochem. Biophys. Methods* **5**, 19.

Laws, W. R., and Brand, L. (1979). *J. Phys. Chem.* **83**, 795.

Leese, R. A., and Wehry, E. L. (1978). *Anal. Chem.* **50**, 1192.

Lewis, C., Ware, W. R., Doemeny, L. J., and Nemzek, T. L. (1973). *Rev. Sci. Instrum.* **44**, 107.

Lin, C.-T., and Sutin, N. (1976). *J. Phys. Chem.* **80**, 97.

Love, J. C., and Demas, J. N. (1983a). *Anal. Chem.* **55**.

Love, J. C., and Demas, J. N. (1983b). *Rev. Sci. Instrum.* **53**.

Love, L. J. C., and Shaver, L. A. (1980). *Anal. Chem.* **52**, 154.

Love, L. J. C., and Upton, L. M. (1980). *Anal. Chem.* **52**, 496.

Lowdermilk, W. H. (1979). *In* "Laser Handbook" (M. L. Stitch, ed.), Vol. 3, p. 361. North-Holland Publ., Amsterdam.

Lyke, R. L., and Ware, W. R. (1977). *Rev. Sci. Instrum.* **48**, 320.

Lytle, F. E. (1973). *Photochem. Photobiol.* **17**, 75.

Lytle, F. E. (1974). *Anal. Chem.* **46**, 817A.

Mackey, R. C., Pollack, S. A., and Witte, R. S. (1965). Rep. No. 9807-6004-RU-000. TRW Systems, Redondo Beach, California.

McKinnon, A. E., Szabo, A. G., and Miller, D. R. (1977). *J. Phys. Chem.* **81**, 1564.

Mandal, K., Pearson, T. D. L., and Demas, J. N. (1980). *Anal. Chem.* **52**, 2184.

Mandelis, A., and Royce, B. S. H. (1980). *J. Opt. Soc. Am.* **70**, 474.

Mantulin, W. M., and Weber, G. (1976). *J. Chem. Phys.* **66**, 4092.

Marquardt, D. W. (1963). *J. Soc. Ind. Appl. Math.* **11**, 431.

Matthews, T. G., and Lytle, F. E. (1979). *Anal. Chem.* **51**, 583.

Merkel, P. B., and Kearns, D. R. (1971). *Chem. Phys. Lett.* **12**, 120.

Mitchell, G. W., and Spencer, R. D. (1981). "Application Brief," Vol. 1, No. 1. SLM Instruments, Urbana, Illinois.

Moya, I. (1983). *In* "Time-Resolved Fluorescence Spectroscopy in Biochemistry and Biology" (R. B. Cundall and R. E. Dale, eds.), pp. 741, 755. Plenum, New York.

Munro, I. D. (1983). *In* "Time-Resolved Fluorescence Spectroscopy in Biochemistry and Biology" (R. B. Cundall and R. E. Dale, eds.), p. 81. Plenum, New York.

Murao, T., Yamazaki, I., and Yoshihana, K. (1982). *Appl. Opt.* **21**, 2297.

Nelder, J. A., and Mead, R. (1965). *Comput. J.* **7**, 308.

Nelson, R. (1981). *EDN* June 10, p. 77.

O'Connor, D. V., Ware, W. R., and Andre, J. C. (1979). *J. Phys. Chem.* **83**, 1333.

O'Connor, D. V., Meech, S. R., Phillips, D. (1982). *Chem. Phys. Lett.* **88**, 22.

O'Connor, D. V., Roberts, A. J., Lampert, R. A., Meech, S. R., Chewster, L. A., and Phillips, D. (1983). *Anal. Chem.* **55**, 68.

Ortec, Inc. (1975). "Model 473 Constant Fraction Discriminator," Instrument Manual. Oak Ridge, Tennessee.

Ott, H. W. (1976). "Noise Reduction Techniques in Electronic Systems," Wiley, New York.

Parker, C. A. (1968). "Photoluminescence of Solutions," Elsevier, Amsterdam.

Pearson, T. D. L., Demas, J. N., and Davis, S. (1982). *Anal. Chem.* **54**, 1899.

Peirce, B. O., and Foster, R. M. (1957). "A Short Table of Integrals," 4th ed. Ginn (Blaisdell), Boston, Massachusetts.

Perrin, F. (1929). *Ann. Phys. (Paris)* **12**, 169.

Peterson, S. H., and Demas, J. N. (1976). *J. Am. Chem. Soc.* **98**, 7880.

Peterson, S. H., and Demas, J. N. (1979). *J. Am. Chem. Soc.* **101**, 6571.

Peterson, S. H., Demas, J. N., Kennelly, T., Gafney, H., and Novak, D. P. (1979). *J. Phys. Chem.* **83**, 2991.

Plumb, I. C., Copper, G. H., and Heap, D. G. (1977). *J. Phys. E* **10**, 744.

Rayner, D. M., McKinnon, A. E., Szabo, A. B., and Hackett, P. A. (1976). *Can. J. Chem.* **54**, 3246.

Rayner, D. M., McKinnon, A. E., and Szabo, A. G. (1977). *Rev. Sci. Instrum.* **48**, 1050.

RCA (1980). "Photomultiplier Handbook," Tech. Ser. PT-61. Lancaster, Pennsylvania.

Reed, F. W., and Demas, J. N. (1983). *In* "Time-Resolved Fluorescence Spectroscopy in Biochemistry and Biology" (R. B. Cundall and R. E. Dale, eds.), p. 285. Plenum, New York.

Rockley, M. G. (1980). *Biophys. J.* **30**, 193.

Rodgers, M. A. J. (1981). *Chem. Phys. Lett.* **78**, 509.

Rogers, P. C. (1962). Tech. Rep. No. 76 (NYO-2303). Mass. Inst. Technol., Cambridge, Massachusetts.

Rutherford, E., and Geiger, H. (1910). *Philos. Mag.* **20**, 698.

Ryan, P. B., Barr, R. L., and Todd, H. D. (1980). *Anal. Chem.* **52**, 1460.

Santoni, A. (1980). *EDN* Dec. 15, p. 182.

Scarborough, J. B. (1962). "Numerical Mathematical Analysis," Johns Hopkins Press, Baltimore, Maryland.

Schaap, A. P., ed. (1976). "Singular Molecular Oxygen," Benchmark Papers in Organic Chemistry, Vol. 5. Dowden, Hutchinson & Ross, Stroudsburg, Pennsylvania.

Schäfer, F. P., ed. (1977). "Dye Lasers," 2nd rev. ed. Springer-Verlag, Berlin and New York.

Schubert, V. D., Wabnitz, H., and Wilhemli, B. (1980). *Exp. Tech. Phys.* **28**, 435.

Shank, C. V., and Ippen, E. P. (1977). "Topics in Applied Physics," Vol. 1. 2nd rev. ed., Springer-Verlag, Berlin and New York.

Shavers, C. L., Parsons, M. L., and Deming, S. N. (1979). *J. Chem. Educ.* **56**, 307, 812.

Sholes, R. R., and Small, J. G. (1980). *Rev. Sci. Instrum.* **51**, 882.

Small, E. W., and Isenberg, I. (1976). *Biopolymers* **15**, 1093.

Small, E. W., and Isenberg, I. (1977). *J. Chem. Phys.* **66**, 3347.

Small, E. W., and Isenberg, I. (1983). *In* "Time-Resolved Fluorescence Spectroscopy in Biochemistry and Biology" (R. B. Cundall and R. E. Dale, eds.), p. 199. Plenum, New York.

Spears, K. G., Cramer, L. E., and Hoffland, L. D. (1978). *Rev. Sci. Instrum.* **49**, 255.

Spears, K. G., Steinmetz-Bauer, K. M., and Gray, T. H. (1980). In "Picosecond Phenomena II," p. 106. Springer-Verlag, Berlin and New York.

Spectra-Physics (1981). "Pulsed Laser Systems," Information Sheet. Mountain View, California.

Spencer, R. D., and Weber, G. (1970). *J. Chem. Phys.* **52**, 1654.

Spendley, W., Hext, G. R., and Himsworth, F. R. (1962). *Technometrics* **4**, 441.

Steinberg, I. Z., and Katchalski, E. (1968). *J. Chem. Phys.* **48**, 2404.

Steinberg, I. Z., Haas, E., and Katzir-Katchalski, E. (1983). *In* "Time-Resolved Fluorescence Spectroscopy in Biochemistry and Biology" (R. B. Cundall and R. E. Dale, eds.). Plenum, New York.

Stern, O., and Volmer, M. (1919). *Phys. Z.* **20**, 183.

Strickler, S. J., and Berg, R. A. (1962). *J. Chem. Phys.* **37**, 814.

Stryer, L. (1978). *Annu. Rev. Biochem.* **47**, 819.

Stryer, L., and Haugland, R. P. (1967). *Proc. Natl. Acad. Sci. U.S.A.* **58**, 719.

Suzuki, R., Umezu, K., Takuma, H., and Shimizu, F. (1981). *Rev. Sci. Instrum.* **52**, 287.

Tao, T. (1969). *Biopolymers* **8**, 609.

Taylor, D. G., and Demas, J. N. (1979). *Anal. Chem.* **51**, 712, 717.

Taylor, D. G., Turley, T. J., Rodgers, M. L., Peterson, S. H., and Demas, J. N. (1980). *Rev. Sci. Instrum.* **51**, 855.

Tektronix (1976). "R7912 Interface Concepts: Instruction Manual," Part No. 070-1881-00. Beverton, Oregon.

TRW Systems (1967). "Measurement of Short Decay Times With the TRW Nanosecond Spectral Source System," Appl. Note 6. Redondo Beach, California.

Turley, T. J. (1980). M.S. Thesis, Univ. of Virginia, Charlottesville.

Turro, N. J. (1978). "Modern Molecular Photochemistry." Benjamin, New York.

Valeur, B., and Moirez, J. (1973). *J. Chim. Phys.* **70**, 500.

Veselova, T. V., and Shirokov, V. I. (1972). *Bull. Acad. Sci. USSR, Phys. Ser. (Engl. Transl.)* **36**, 925.

Veselova, T. V., Cherkasov, A. S., and Shirokov, V. I. (1970). *Opt. Spectrosc.* **29**, 617.

Wahl, P. (1975). "New Techniques in Biophysics and Cell Biology." Wiley, New York.

Wahl, P. (1983). *In* "Time-Resolved Fluorescence Spectroscopy in Biochemistry and Biology" (R. B. Cundall and R. E. Dale, eds.), pp. 483, 497. Plenum, New York.

Wahl, P., Auchet, J. C., and Donzel, B. (1974). *Rev. Sci. Instrum.* **45**, 28.

Wallenstein, R. (1979). "Laser Handbook," (M. L. Stitch, ed.), p. 289. North-Holland Publ., Amsterdam.

Ware, W. R. (1971). *In* "Creation and Detection of the Excited State" (E. LaMola, ed.), Vol. 1, Part A, p. 213. Dekker, New York.

Ware, W. R. (1983). *In* "Time-Resolved Fluorescence Spectroscopy in Biochemistry and Biology" (R. B. Cundall and R. E. Dale, eds.), p. 23. Plenum, New York.

Ware, W. R., Doemeny, L. J., and Nemzek, T. L. (1973). *J. Phys. Chem.* **77**, 2038.

Watts, R. J., and Bergeron, S. F. (1978). *J. Phys. Chem.* **83**, 424.

West, M. A. (1976). *In* "Creation and Detection of the Excited State" (W. R. Ware, ed.), Vol. 4, p. 217. Dekker, New York.

Wiesenfeld, J. M., Ippen, E. P., Corin, A., and Bersohn, R. (1980). *J. Am. Chem. Soc.* **102**, 7256.

Wild, U., Holzwarth, A., and Good, H. P. (1977). *Rev. Sci. Instrum.* **48**, 1621.

Winefordner, J. D. (1973). *In* "Accuracy in Spectrophotometry and Luminescence Measurements" (R. Mavrodineanu, J. I. Schultz, and D. Menis, eds.), Spec. Publ. No. 378, p. 169. Natl. Bur. Stand., Washington, D.C.

Wright, G. T. (1955). *Proc. Phys. Soc. London, Sect. B* **68**, 241.

Wyker, A. T., and Demas, J. N. (1976). *J. Chem. Educ.* **53**, 656.

Young, R. H., Brewer, D., and Keller, R. (1973). *J. Am. Chem. Soc.* **95**, 375.

Zimmerman, H. E., Werthemann, D. P., and Kamm, K. S. (1974). *J. Am. Chem. Soc.* **96**, 439.

Zimmerman, H. E., Goldman, T. D., Hirzel, T. K., and Schmidt, S. P. (1980). *J. Org. Chem.* **45**, 3933.

Index